# Why Can't Elephants Jump?

**AND 113 OTHER TANTALIZING
SCIENCE QUESTIONS ANSWERED**

# Why Can't Elephants Jump?

## AND 113 OTHER TANTALIZING SCIENCE QUESTIONS ANSWERED

## NewScientist

PEGASUS BOOKS
NEW YORK

WHY CAN'T ELEPHANTS JUMP?

Pegasus Books LLC
80 Broad Street, 5th Floor
New York, NY 10004

First Pegasus Books edition 2011

Library of Congress Cataloging-in-Publication Data is available.

ISBN: 978-1-60598-261-8

10 9 8 7 6 5 4 3 2 1

Printed in the United States of America
Distributed by W. W. Norton & Company, Inc.
www.pegasusbooks.us

# Contents

Introduction                                    1

1   Food and drink                              3

2   Our bodies                                 28

3   Domestic science                           78

4   Plants and animals                        104

5   Our planet, our solar system              153

6   Weird weather                             175

7   Troublesome transport                     189

8   Best of the rest                          203

    Index                                     224

# Introduction

We're back. And even better than before. This time we're going to tell you why wet things smell more than dry things and we'll answer that age-old question on everybody's lips: how wise is it to pick your nose and subsequently consume what you find in there. Yes, welcome to *Why Can't Elephants Jump?*, the latest book in a series that began with *Does Anything Eat Wasps?* and had its last outing more than two years ago with *Do Polar Bears Get Lonely?* In this edition we'll reveal whether it's possible to see the curvature of the Earth from anywhere on its surface (even atop Blackpool Tower) and we'll explain which is more tiring, walking up a slope or climbing steps.

Most of these questions began life on the Last Word page of *New Scientist* magazine. Each week readers pose everyday science questions while others attempt to answer them. You can join them by buying the magazine or visiting the website at www.last-word.com. But remember, you won't find anything on quantum mechanics or event horizons – that's all covered by the clever *New Scientist* journalists elsewhere on the website and in the magazine. Instead, you'll find the science of everyday things, such as why it's difficult to tear sticky tape across its width. New questions are always welcomed, as is hefty debate on whether we gave the right answers to old ones. And our knowledge is growing at a rapid rate. Once upon a time we worried that the observational prowess and ingenuity of our readers would dry up – now we know better, and this book is testimony to those powers.

Eagle-eyed readers will have noticed that a good few of the questions in this book – just like those found in its predecessors – have been answered by a Mr Jon Richfield of Somerset West, South Africa. In fact, Jon's existence has been questioned by many (some speculated he was a pseudonymous stand-in for *New Scientist*'s journalists), while others have wanted to know if there is anything he cannot answer. We are delighted both to confirm his existence and to run a question that his wife tells us he doesn't know the answer to. Check out page 51.

And last of all, we are daring to go into the casino bar once more with James Bond. In previous books we've mused over what chemistry creates the difference between a shaken and a stirred vodka martini. Well, we've found out more. You expect us to talk? Well, if you insist on aiming that scary-looking laser at us, of course we will. Turn to page 6.

Enjoy the book and be inspired to join in. You could soon be the next Jon Richfield.

Mick O'Hare

Special thanks are due to everyone at Profile books – especially Paul Forty and Valentina Zanca – while Jeremy Webb, Jessica Griggs, Lucy Dodwell and the subbing team at *New Scientist* made sure the book is as error-free as it could be – any remaining mistakes are mine alone. Judith Hurrell proved to be a tireless researcher when press-ganged into the job unexpectedly, so thanks to her. And finally thanks indeed to Sally and Thomas.

# 1 Food and drink

## ? Super food

*Is there a single foodstuff that could provide all the nutrients that a human needs to stay reasonably healthy indefinitely?*

**Andy Taplin**
Cambridge, UK

Any single substance such as water or fat? No. Any single tissue such as muscle or potato? No. But if we allow free drinking water and air to breathe – even though those are also nutrients – then we can relax our rules. Even so, drinking milk while eating corn would count as two foodstuffs, and who knows how many foodstuffs pizza contains.

Not surprisingly, no strict monodiet can rival any healthily balanced diet, but there are two classes of foodstuffs that in appropriate quantities can maintain a reasonable level of health. One such class is baby food. Examples include eggs, milk, certain seeds, and so on. None is a perfect option, but some are adequate.

Alternatively one might cheat by counting essentially whole animals: oysters or fish such as whitebait or sardines might supply the necessary nutrient uptake. Animals sufficiently closely related to humans might also do, if eaten in the correct form and quantity. Farming families in the semi-desert Karoo region of South Africa apparently ate mainly sheep or cattle.

For the most perfectly balanced human monodiet, however, other humans would be the logical food of choice. Not sure there would be many takers though.

**Jon Richfield**
Somerset West, South Africa

Despite various claims made over the years for spinach, baked beans or bananas, the answer is 'no'. To remain healthy over the course of a natural lifetime, a human needs a balanced diet, with a combination of carbohydrates and protein and a proper range of vitamins. The balance may vary with age, from individual to individual, and even between societies in differing environments, but balance is the key to healthy eating.

Probably the most homogenous diet of any human community is that of the Inuit in Arctic North America, which traditionally consists of 90 per cent meat and fish and effectively no carbohydrates. The explorer Vilhjalmur Stefansson not only established that Inuit hunters often lived for between six and nine months of the year on a wholly carnivorous diet, but also claimed to have sustained himself by eating just meat and fish on his expeditions.

In a series of controlled experiments under the auspices of *The Journal of the American Medical Association*, Stefansson and a number of his colleagues reproduced the dietary regime they had followed in the Arctic, without any apparent ill effects and without, much to the supervising doctors' surprise, developing scurvy. However, subsequent long-term studies of health factors in the Inuit community have established a strong correlation between the carnivorous diet and early deaths among Inuit men from heart attacks and other cardiac problems.

The bottom line is to remember what your mother told you: always eat your greens.

**Hadrian Jeffs**
Norwich, Norfolk, UK

Humans have been suggested as the ideal diet for other humans, but eating human flesh would not satisfy all our dietary needs, no matter how healthy the diner or victim, for a very simple reason: not every nutrient, mineral and vitamin in the body remains available to the next step in the food chain. Some substances are 'used up' and others are built into indigestible tissues and structures.

Cooking can increase the digestibility of many foodstuffs for a human, but things like hair, bones and teeth cannot be prepared to make them amenable to our digestion. However, we need the minerals, amino acids and other nutrients in them to make these substances for ourselves.

The human body is a tenacious machine and will continue to survive on a very poor diet for quite a while. Many people living in harsh climates have their own dietary supplements in local foods and 'delicacies' that they may not even realise are making up a shortfall. The same is true of those who live in a self-imposed harsh regime, such as true vegetarians. One can live on such diets and remain healthy as long as there is a balanced variety of nutrients, or by taking artificial supplements such as vitamin B12.

Some natural foodstuffs provide a reasonable balance of necessary nutrients, but the only known and proven foodstuffs that truly provide everything that a human body would need, in a single wrapper, are manufactured. Survival rations and trail foods are many and varied, but remain unpopular because they are generally dry and unpalatable.

Foods made to cover all the needs of large and physiologically similar mammals, such as dogs, pigs and other omnivores, could probably sustain us very well too, although we may have to eat more of the stuff than we'd like to get sufficient supplies of any human-specific nutrients.

**Nat Guthrie**
By email, no address supplied

## ❓ Stirring stuff

*What is the significance of James Bond's famous phrase 'shaken, not stirred'? Is there really a difference in the taste of a shaken vodka martini, as opposed to a stirred one? And if so, why?*

**Mark Langford**
Stockport, Cheshire, UK

*The dispute over the difference between a shaken or stirred martini has run for some years in the Last Word. An earlier book in this series,* How to Fossilise Your Hamster, *drew on the following three answers to explain the differences. However, more information has come to light – Ed.*

Supposedly, when a martini is shaken, not stirred, it 'bruises' the spirit in the martini. To seasoned martini drinkers this changes the taste.

**Padraic O'Neile**
Newcastle, New South Wales, Australia

Because a martini is to be drunk within seconds of preparation rather than minutes, there is a difference. The tiny bubbles caused by the shaking mean that a well-shaken martini is cloudy. This will also have an effect on the texture of the drink – it is less oily than the stirred version – hence the taste. The long-standing assumption that the spirit is bruised by the process is nonsense because vodka does not have a vascular system.

**Peter Brooks**
Bristol, UK

James Bond may have appreciated the softening and ripening effect of partial oxidation of the aldehydes in vermouth – akin to letting red wine breathe before drinking. In a refined and homogeneous substrate such as vodka martini, a good shake can speed the process.

**Alan Calverd**
Bishops Stortford, Hertfordshire, UK

*We have since learned, however, that other chemical reactions may be taking place – Ed.*

Biochemists at the University of Western Ontario in Canada have suggested that the change in flavour is not caused by the oxidation of aldehydes, but because shaking martinis can break down hydrogen peroxide present in the drink. Stirred martinis have double the amount of hydrogen peroxide that shaken ones contain. This significantly affects the flavour.

**Peter McNally**
Vancouver, British Columbia, Canada

The reason that shaken martinis are cloudy is not so much down to bubble formation from the cocktail-shaking process but rather that the crushed ice used in the shaker deposits tiny crystals into the poured drink. These make the drink more cloudy and then slowly melt, allowing it to clear.

**Frank Melly**
New York City, US

*This clearly called for more research. Was it bubbles or ice that caused the cloudiness in a shaken martini? And could either account for any difference in taste? First up we needed a good martini recipe. This one came from cocktail mixologist Eric Keitt, who works the bar at Oceanaire in Washington DC:*

> *Double vodka and a few drops of dry vermouth*
> *Pour into a cocktail shaker with crushed ice*

*Shake until the hand holding the shaker is very cold*
*Strain into a martini glass*
*Add an olive or a twist of lime zest*

*Eric tells us: 'The application of vermouth should be a few – and I mean a few – drops, maybe two or three. Vermouth will release the aromatics of the vodka, making for a more enjoyable drink.' Eric is a fan of stirring the drink for the reasons given below, but in this instance we had no choice but to shake.*

*Three martinis were prepared. The first was shaken with crushed ice. The drink was very cloudy and took a long time to clear but, as far as we could ascertain, the cloudiness was caused only by tiny bubbles from the shaking plus the condensation on the chilled martini glass. No ice crystals were present unless they were microscopic.*

*The second was a room-temperature martini, shaken without ice. Bubbles formed in this when it was poured but quickly dissipated, much faster than in the iced martini.*

*The third martini was made in an attempt to replicate the chilled conditions of the iced martini but without adding ice to the shaker. The martini and its shaker were wrapped in a drinks chiller until ice cold and the same temperature as the first martini. Then it was shaken. When poured it was cloudy for much longer than the room-temperature martini but not for as long as the iced martini.*

*From this we ascertained that the ice does have some effect on the clouding process, as do cold conditions. Iced martinis do produce the cloudiest drink, but no ice spicules appeared present in the drink, contrary to the suggestion above by Frank Melly. Chilled martinis without ice produced a cloudy drink too, but for a shorter time than iced martinis. The room-temperature martini cleared the fastest. So nothing conclusive here yet, except that temperature plays a role of some kind – more experimentation is needed. Is there any reader out there who can examine the shaken martini microscopically to rule out (or, indeed, confirm) the presence of ice crystals?*

*But there's more. Putting cloudiness to one side for the moment, Anna Collins seems to have answered the original question of what makes a shaken martini taste different to a stirred one, her suggestion apparently being confirmed in a blind tasting – Ed.*

The reason Bond orders his martinis shaken is that the ice helps to dissipate any residual oil left over in the manufacture of vodka from potatoes – the base vegetable for many vodkas at the time Ian Fleming's original novels were written. With the rise of higher-quality grain vodkas the process is now unnecessary. In fact, shaking the martini with ice dilutes it too much for many fans of the drink. Stirring chills the martini without losing its essential strength.

**Anna Collins**
Washington DC, US

Anna Collins is correct, according to our blind trial. We bought two bottles of vodka, one grain based, the other potato based. First we tasted the vodkas. In the blind trial all six people in our sample said the potato vodka was oily, the grain vodka wasn't. Then we made two vodka martinis using the potato vodka. One was stirred with ice, the other shaken with ice. The difference was quite distinct and in a blind tasting every one of the six drinkers correctly identified the shaken martini as being much less oily. But the martini had to be consumed quickly. If left to settle for about 5 minutes or so, the shaken martini became oily again.

**Peter Simmons**
London, UK

*Maybe that's the last word on vodka martinis. Although knowing our readers' propensity to keep unearthing new evidence, we suspect not – Ed.*

# ❓ Stick that!

*Having discovered the joys of the 'appletini' (vodka mixed with apple juice, cider or apple liqueur), I have a question. The garnish is a slice of apple and a Maraschino or glacé cherry on a cocktail stick. If the cherry is at the bottom of the stick it floats in the appletini, but with the apple slice at the bottom it sinks. Why? Surely the buoyancy of the two items combined is an absolute and their orientation should make no difference. I shall leave it to the imagination as to the number of 'tinis consumed before this anomaly became a burning topic of conversation.*

**Richard Batho**
St Saviour, Jersey

Ethanol acts as a wetting agent, so in an alcoholic drink the submerged slice of apple holds too little air to float the non-buoyant, sugary cherry. The assembly will thus sink, though if you add soda, enough bubbles might attach to the apple to make the whole thing float again.

Human sloshing complicates insights after the fourth glass of appletini, but buoyancy is a more complicated affair than density considerations might suggest. For example, a boat that is seaworthy might sink if capsized.

Try dropping a clean, dry pin or razor blade gently onto a glass of still, pure water. Dropped endwise the object sinks; the metal is too dense. But surface tension will support the item if you gently drop it flat onto the fluid, especially if the metal is lightly waxed or oiled.

The behaviour of your appletini garnish is similar in some ways. The waxy skin and the broad shape of an unpeeled slice of apple on the surface of the drink can resist both the wetting and the shipping of fluid over the edge of the slice.

The stimulation of considering this question's complications should mitigate the brain-addling aspects of appletini, though, sadly, not to the extent of fully reversing them.

**Antony David**
London, UK

Surface tension effects probably account for the observations. If the buoyancy of the cocktail stick assembly is nearly neutral, a flat slice of apple, when uppermost and level with the cocktail surface, may provide a sufficiently long perimeter for surface tension to hold the assembly up. The spherical cherry in the same position has little or no perimeter on which surface tension can act. The coating on the cherry may also reduce surface tension.

Try replacing the apple slice with a small ball of apple and see if the stick now sinks even with the apple uppermost. If this experiment fails, the cocktail may have aged, so drink it and try again with a new one...

**Paul Gladwell**
Northwich, Cheshire, UK

#  Hot to trot

*Mustard and chillies are both hot, but the burning sensation from a chilli stays in the mouth for ages while the sensation from hot mustard disappears in a few seconds. Why is this?*

**Dominic Lopez-Real**
By email, no address supplied

The chemical mainly responsible for the burning spice in chilli peppers is capsaicin, a complicated organic compound that binds to receptors in your mouth and throat, producing the desired (or dreaded) sensation.

Capsaicin is an oil, almost completely insoluble in water. This is why you need a fat-containing substance like milk to wash it away – watery saliva doesn't do the trick.

On the other hand, the compound responsible for the hotness of mustard (as well as horseradish and wasabi) is called allyl isothiocyanate. This chemical is slightly water-soluble, and can be more readily washed away into the stomach by saliva.

Further, the chemical in mustard is more volatile than capsaicin so it evaporates more readily, allowing its fumes to enter the nasal passages (explaining why the burning sensation from mustard is often felt in the nose). These fumes can be easily removed by breathing deeply, a useful strategy if the sensation becomes overwhelming.

**Zachary Vernon**
Toronto, Canada


The hotness of mustard comes from allyl isothiocyanate, which is formed when myrosinase and sinigrin (in mustard seeds) react together in water. It dissolves well in most organic compounds, and to an extent in water, and is also volatile, so will quickly disperse.

On the other hand, capsaicin, the hot ingredient of chillies, is not very water-soluble. So its heat tends to stay. It is soluble, however, in alcohol, which raises the question: which came first, the lager or the vindaloo?

**Matt Billingham**
Switzerland

# ? Fungal goodness

*We are constantly being exhorted to eat five servings of fruit
or vegetables a day, cut down on red meat, eat more fish and so
on. But very little of this advice mentions that other kingdom of
gastronomic delights, fungi. What nutritional value does your
average edible fungus have?*

**Rachel Cave**
Galway, Ireland

Until recently the village of Bourré in central France, where I
live, was a major production centre for mushrooms. Now all
we have left is an artisanal operation as a tourist attraction.

However, the two kinds of mushrooms that were the
mainstays of the industry are still produced here: *Agaricus
bisporus*, or champignon de Paris, which in English tends
to be simply called a 'mushroom'; and *Lentinula edodes*, the
shiitake mushroom. The former has slightly more than 3
grams of protein per 100 grams, and a range of trace minerals
including calcium, iron, magnesium, phosphorus, potassium,
zinc, copper and manganese. It also contains vitamin C and
several B vitamins. Shiitake mushrooms contain rather more
zinc but are lower in protein and vitamin C.

As a vegetarian I find mushrooms invaluable as a way
of providing some variation in the texture of my food. They
tend to go well with many sauces usually designed for meat.
Nutritionally, they compare reasonably well with other foods
from non-animal sources. And cooked properly, in olive oil
with garlic and thyme, for example, they taste great.

**Steve McGiffen**
Bourré, Loir-et-Cher, France

Fungi, mostly represented by mushrooms and truffles, are
edible, nutritionally rich organisms. Mushrooms are an

excellent source of proteins, minerals and dietary fibre, with only small quantities of fats, cholesterol and fatty acids. They are also a source of three essential B vitamins – riboflavin, niacin and pantothenic acid – along with other vitamin groups. All told, mushrooms make an exceptionally good foodstuff for people with diabetes or high cholesterol.

Several Basidiomycetes species have been reported to contain phytochemicals that might be beneficial in the fields of immunology and cardiology, while preliminary results on the effect of lectin from the common edible mushroom, *Agaricus bisporus*, have shown some potential for treating psoriasis.

**Saikat Basu**
Lethbridge, Alberta, Canada

# Cool colour

*Why is frozen milk yellow?*

**Mickey Wright**
Armley, West Yorkshire, UK

The yellow colour of frozen milk comes from the vitamin riboflavin, which actually got its name from its colour – flavus is the Latin for yellow.

Riboflavin is dissolved in the watery portion of milk, which is also filled with minute particles of protein and droplets of butterfat. In fresh milk, all the suspended particles and droplets scatter any light that strikes them evenly, so that the milk appears opaque and white – milky, in other words.

However, as the milk freezes and most of its water crystallises into ice before other substances, the normally dilute riboflavin becomes concentrated in the remaining liquid water.

This means these areas start to turn yellow and, as the clear water-ice crystals form, we are able to see it.

**Harold McGee**
San Francisco, California, US

*Harold McGee is the author of* On Food & Cooking: The science and lore of the kitchen *(Fireside, 1997) – Ed.*

# ? Down in one

*Why is it so much easier to drink a whole pint of beer or orange squash, say, in one go than it is to down a pint of water?*

**Geoff Lane**
Bury, Lancashire, UK

Is it fair to suggest that the questioner's preference for downing a pint of beer rather than a pint of water is dependent on individual taste?

I have never been able to down a whole pint of beer, much to my dismay – and not for want of trying! But I am able to drink a whole pint of water or more in one go, even if I do feel quite sick afterwards. I much prefer the taste of water to beer and I find it difficult to consume any fizzy liquids in large quantities. I assumed that this was the case for most people, but obviously not.

**Jayne Staines**
Eastbourne, East Sussex, UK

Is this just hearsay or is it the result of a well-designed experiment? I suspect the former.

Even so, a possible explanation could be that orange squash and beer have strong and pleasant flavours while water is bland, and that there is more motivation in a pub to persist with beer or orange squash than with water.

But surely it depends on context: if I were seriously dehydrated after some time in a desert, say, I would certainly prefer a pint of water to a pint of beer. In a pub, where one is not dehydrated, beer would be my preference.

So we need to establish whether the questioner's proposition is really true (the clincher would be for non-drinkers to compare beer with another carbonated drink, such as cola) and whether it is true in a variety of contexts. If there is experimental evidence, then we can examine it and see if it was well designed and worth believing. Then we would need to find out why. This is the correct scientific response.

**Mike O'Mahony**
Professor and sensory scientist
Department of Food Science and Technology
University of California, Davis, US

## ? Veggie violence

*Occasionally, when reheating broccoli and sweet potato in the microwave, what sounds like violent electrical arcing occurs if the two are in contact with each other. This results in blackened sections. What is going on?*

**Stephen Plevier**
Hadfield, Victoria, Australia

One may think of microwaves in the oven as light shining through translucent food, with the light that gets absorbed being turned into heat. This is a good description of what happens when the food is a continuous mass measuring more than about 6 centimetres in all directions.

However, if the food mass is smaller and oddly shaped, the interaction between the wavelength of the radiation (usually around 12.24 centimetres) and the shape becomes dominant. Microwaves then behave more like radio waves, and some slender, spiky or complex-shaped vegetables become electrical conductors. These conductors act like radio aerials along which electric charges surge back and forth, typically at about 2.5 billion times a second. Wherever these 'aerials' form small gaps or fine contacts, electrons leap across, creating hot spots or sparks that scorch the food. This is why conductors such as forks, plates decorated with metal leaf, or unsuitably shaped foods cause arcing.

If your broccoli is in firm contact with the lumps of sweet potato or thoroughly wet, the fluid clogs the gaps and masks any fine tips, preventing harm, but wherever the electrons can arc across fine contacts or gaps, you get sparks or charring.

**Brian Allen-Smith**
Philadelphia, Pennsylvania, US

# ? Weighty flier

*As part of Christmas dinner this year I cooked a tasty goose. I was astounded at the amount of fat that poured off it during cooking. Why do geese need so much fat?*

**Debora MacKenzie**
Brussels, Belgium

I would argue that geese don't need this fat, but rather that their intensive rearing has provided them with an excess of calories which they have laid down as fat, just as occurs in many other animals – including humans and their pets if they overeat and under-exercise.

Wild geese have very lean carcasses because their diet typically comprises grasses which contain small amounts of energy, on which they graze for long periods each day and which they often have to fly long distances to find.

Farm-reared geese, on the other hand, though they may actually be free-range, will have access to high-energy concentrated diets similar to those of broiler chickens. They expend very little energy each day on their basic functions and so will store the majority of calories as fat once they have achieved their mature size.

**Ian Jeffcoate**
University of Glasgow Veterinary School
Glasgow, UK

In the wild, geese are aquatic birds, with many species being migratory. They consequently need both a substantial energy reserve to sustain them during the long flights of their migration, and good insulation to protect them from the cold and wet. Fat covers both necessities quite nicely.

Goose fat, therefore, is by no means simply dead weight. An adult greylag (the species from which almost all breeds of domesticated goose descend) can weigh up to 5.5 kilograms. With a wingspan of over 160 centimetres, that makes for a relatively low wing loading – the ratio of mass to wing area.

In aviation terms, geese could be considered the 'long-haul wide-bodies' of the bird world, and a fuel load that can sustain them over thousands of kilometres is vital rather than a burden.

Of course, the ratio of goose fat to body weight is much lower in a wild bird than in the domesticated fowl typically eaten for Christmas dinner. For a number of reasons, domestic geese have had their body fat augmented by a combination of selective breeding and diet.

Prior to the introduction of the railways, from a farmer's

perspective the goose had quite a big advantage over the turkey – it was a lot easier to transport. Goose farmers based in East Anglia in the UK, for example, would simply walk their geese to London's Smithfield market. They knew that, unlike turkeys, a goose could not roost overnight up a tree from which it would be almost impossible to retrieve the following morning.

Furthermore, geese could sustain themselves on this 160-kilometre waddle by grazing on grass, drawing on their reserves of fat for additional sustenance.

From a cook's perspective, the goose's advantage over the turkey was that it required no additional basting, its own supply of body fat being sufficient if spooned over the bird periodically while roasting. Indeed, so much fat comes off a full-grown bird that the excess can be used to baste other meat cooked at the same time, a favourite culinary trick of Charles Dickens.

Through his writing, Dickens helped make turkey an essential part of the British Christmas dinner. Ironically, he preferred the taste of goose, and to further his enjoyment of the bird as part of his seasonal lunch he would roast a whole ox heart in a pan placed underneath the trivet on which the goose was cooking, so that the heart's bland but succulent meat would soak up the goose's flavour, along with its fat as it dripped from above.

**Hadrian Jeffs**
Norwich, Norfolk, UK

## ❓ Spice attack

*An earlier book in this series told us why garlic makes your breath and body smell, but I want to know why the spice methi, or fresh fenugreek, has a similar, possibly stronger, effect.*

**Nikki Bedi**
BBC Radio Asian Network, UK

Depending on their biochemical nature, volatile components of foods or their metabolic products enter the blood and exit via lungs, urine, sweat, saliva or sebum, more than most people notice. As a result, families or communities with distinctive cuisines have distinct body scents.

There are many examples beyond obvious ones such as asparagus and onion-like foods. Stewed mutton and beef give a recognisable odour to one's urine. No doubt any self-respecting dog could identify other meats.

Many nitrogen compounds are particularly likely to be excreted in urine or sweat. I rather like the yeasty smell of thiamine, but my wife hates it, as does a friend who once had to have daily thiamine injections. His skin would reek before the doctor even finished the injection. Some people can even guess which cheese you have eaten in the last few days; presumably the smell gets into your sebum.

Fenugreek contains a range of sulphur and nitrogen-rich aroma molecules that the body modifies and excretes in breath and sweat, but the main burnt-sugar smell comes from the lactone sotolon, whose smell we can detect even in minute concentrations.

**Antony David**
London, UK

# ❓ Con gas, sin gas

*Is fizzy water lighter than still?*

**Asked by listeners on BBC Three Counties Radio, UK**

Fizzy or carbonated water is heavier – that is, denser – than non-carbonated water if the carbon dioxide is in solution rather than forming bubbles. For a given weight of water and dissolved carbon dioxide, the volume can be calculated by adding the volume of the water alone (about 1 millilitre per gram of water) to that of the gas (about 0.8 millilitres for each gram of carbon dioxide). So a solution of, say, 2 grams of carbon dioxide in 998 grams of water would have a mass of 1 kilogram, a volume of 999.6 millilitres and a density of 1.0004 grams per millilitre (at 4 °C, the temperature at which water is at its most dense). There is more information on this at the University of California's eScholarship website (bit.ly/3kxoIU). However, if the water is fizzing, then it will be lighter – less dense – than still water. It is the bubbles that reduce the density: 2 grams of carbon dioxide gas would have a volume of about 1 litre.

Lake Nyos in Cameroon has carbon dioxide seeping into it from below. Usually the carbon dioxide-laden water stays at the bottom of the lake because it is denser, but sometimes it begins to form bubbles which cause it to rise. This sets in train a process that brings up a massive amount of carbon dioxide, which bubbles into the air.

In 1986, 1,700 people were killed by carbon dioxide that escaped from the lake and smothered the surrounding valley (*New Scientist*, 24 March 2001, p. 36).

**Eric Kvaalen**
La Courneuve, France

Fizzy water can be lighter than still water, but it depends on what you define as still water. For example, is the questioner asking about pure distilled water, tap water or mineral water? All these have different densities. Tap water varies a great deal: it is hard and chalky in the south of England, but soft in both Scotland and northern England, for example.

If you assume that still water means pure distilled water, then at atmospheric pressure carbonated water will be denser, but only very slightly so, thanks to the extra weight of the gas dissolved in it. Of course, this assumes that the carbonated water contains no bubbles. If it is fizzing then the bubbles lower the density to below that of still water.

However, if you are considering still mineral water or any water containing dissolved salts, such as some tap water supplies or seawater, then it is difficult to determine the density without proper testing.

For example, seawater has a density of about 1.025 kilograms per litre – greater than that of fresh water. So still mineral water, depending upon where it is from, can be either heavier or lighter than carbonated water.

**Mogg**
By email, no address supplied

# ❓ Citric surprise

*Some friends and I were drinking from a jug of water that contained wedges of both lime and lemon. All the lemon wedges were floating, but all the lime wedges had sunk to the bottom of the jug. There were enough pieces of both for us to infer this was not just coincidence, and all of us were pretty certain that we'd seen lime slices floating before. Can anyone offer an explanation?*

**Bladon Mooney**
Leicester, UK

*Head for this great link to a page on the Steve Spangler Science site (bit.ly/aMMLze), where the answer is tested – Ed.*

This is down to the interplay of two factors: air and solutes. The cells in fruit tissue typically have quite high concentrations of solutes, mainly organic acids for citrus and sugars for apples. In some types this can amount to as much as 18 per cent of the total weight. The more concentrated the solutes, the denser the cells will be and the more likely it is that the fruit will sink.

In plant tissue there are also air spaces between the cells, so the tissue is less dense than the cells are. If the air spaces are large enough, the tissue will float even if the cells alone would be dense enough to sink. Air spaces can range from as little as 1.5 per cent of the tissue's volume in the case of a potato to more than 20 per cent for some leaves.

The effect of air spaces tends to outweigh cell sugar content, which is why a potato sinks while an apple with 15 to 20 per cent air spaces by volume will float, despite the high sugar content of its cells.

A variety of factors affect the volume of the air spaces and concentration of solutes in different fruit, including growing conditions, ripeness and storage conditions. With citrus fruit, a major factor is the peel. The inner white layer – the albedo or pith – is low in solutes and notably high in air spaces, while the edible segments are high in sugar/acid content, and low in air spaces. Peel a mandarin and put pieces of peel and segments in a bowl of water, and you will find the segments sink while the peel floats. Some lemons have a thick peel and are resolute floaters, while holding lemons in storage for a time reduces the thickness of the albedo, making the fruit more likely to sink.

So, more than anything, what determines whether your citrus slice floats or sinks in your gin and tonic is the thickness of the albedo.

**Rod Bieleski**
Retired plant physiologist
Devonport, New Zealand

## ❓ Pepper pot

*When a pepper, or capsicum, is cut open there is a space inside, but there are no gaps in the pepper where air could get through. What is the composition of gases in this space, and how did they get there? If a green pepper contains chloroplasts would there be more oxygen and less carbon dioxide in a green pepper than in a red, yellow or orange one?*

**Rosa Clements**
Harrogate, North Yorkshire, UK

The questioner is not entirely correct in suggesting that 'there are no gaps in the pepper where air could get through'. Like most other plant surfaces the surface of a pepper or capsicum has stomata. These are orifices which are controlled by a pair of special cells, the guard cells, to open or shut as the plant requires.

They communicate with an extensive network of air spaces within the tissues, without which the gas exchange required for both photosynthesis and respiration could not take place.

The source of the air is, therefore, the atmosphere, via the stomata and the intercellular air spaces in the wall of the fruit. All capsicum fruit are initially green, with functional chloroplasts, so it is possible that there could be some enrichment in oxygen from photosynthesis at this stage, but not very much, because without gas exchange with the outside air there would be no source of carbon dioxide for further photosynthesis.

When a capsicum ripens to a red or yellow colour the chloroplasts cease to function and turn into chromoplasts containing fibrous deposits of carotenoids and protein. At this stage photosynthesis has ceased and the internal gas will be unlikely to differ very much from the outside air other than in water vapour content.

**Guy Cox**
Sydney, Australia

As the pepper develops, the gases from the atmosphere diffuse into the capsicum's growing cavity. The composition of the internal gases will depend on the respiration rate of the pepper as well as the resistance to gas diffusion. Generally, the more immature the pepper, the higher the respiration rate of the tissues.

We decided to test the internal gas composition of different coloured peppers in our laboratory using gas chromatography. The mean percentage levels of oxygen and carbon dioxide respectively were: green (19.85, 0.068), yellow (18.45, 1.08), red (18.36, 1.15).

It is possible that the higher oxygen/lower carbon dioxide levels in the green fruit were due to photosynthesis because the lab bench was in bright sunshine during all the measurements.

Normally, however, internal light levels are much too low to support any significant photosynthesis in harvested produce.

**Julia Aked and Allen Hilton**
Silsoe, Bedfordshire, UK

## ❓ Ordering a meal

*When we are really hungry the quickest way for our body to gain energy is surely found in the kinds of simple sugars available in desserts. So why is it that we usually eat savoury foods followed by sugary puddings and not the other way around? I have noticed that on aircraft, where all the food arrives together on a tray, people often eat the courses in a quite different order from usual.*

**Toni Lluddmecy**
London, UK

Meals comprising a sequence of courses are arguably unique to civilised humanity. Most animals eat what they can when they can. Given a choice, they proceed from their favourite items to the necessary evils. The more prized the food, the higher the probability of losing it through procrastination and the greater the penalties. Availability is more important than quick energy, so natural selection favours an eye for the main chance, in every species from microbes to hunter-gatherers. Even today, children will go for the goodies first when they can.

Only once humans had achieved security, productivity and the leisure time for multiple-course meals did they formulate the principle of 'never a sweet before the meat'. Long before people discovered that flooding the blood with sugar is an unhealthy habit, they had learned that sweet starters spoil the appetite. This is useful if one needs to discourage guests from overeating, but spoils a good feed and forfeits that pleasant anticipation of treats to follow. Sugary desserts enhance the sense of comfortable repletion and are less harmful at the end of a meal, because diluting the sugar in a gutful of chyme buffers the surge of sugar into the blood.

**Stuart van Dyck**
The Hague, The Netherlands

It's only in relatively modern European cuisine that the sugar is reserved for the last course of the meal. In most of the world, sugar is added to meat dishes, such as in oriental sweet-and-sour recipes or in Mexican mole.

European cuisine did the same until the 17th century, working under a theory of nutrition – inherited from ancient Greece – that sugar was the perfect food. Many meat dishes were sweetened until a new theory developed which saw sugar as harmful and relegated it to a small course served after the main meal, when appetites were lower.

We can still see a remnant of the older cuisine in condiments like steak sauce and ketchup, with their high sugar content.

For more history of this change in European eating habits, search out Rachel Laudan's article 'Birth of the modern diet' in the August 2000 issue of *Scientific American*. Elsewhere, Laudan has pointed out that many supposedly traditional ethnic dishes are less than 100 years old, and created to please tourists.

**Alan Chattaway**
Surrey, British Columbia, Canada

# 2 Our bodies

## ? Double trouble

*Our daughter Aisling would like to know why we have evolved two bodily systems to excrete waste products. Why do we have to both poo and wee?*

**Maeve and Chris Tierney**
Edinburgh, UK

Strictly speaking, the question is misplaced. We do not 'excrete' faeces because our bodies are long fleshy tubes which can be thought of as extremely elongated doughnuts. In a doughnut, one would not consider the hole to be 'inside' the cake. Similarly, the tube from our mouth through our gut to our anus is technically 'outside' the living body.

The process of 'excretion' is the passing of material from inside our bodies to the outside. Our kidneys excrete urine, our skin excretes sweat, our lungs excrete water and carbon dioxide, and the inside of our bowel tube excretes many things along its length to assist digestion, as well as disposing of waste products in our bile. The other excreta our bodies produce include tears, earwax, and various secretions associated with our reproductive processes. If young Aisling suffers (or is about to suffer) from spots, then these too are caused by excretions which have gone awry.

Our faeces, on the other hand, consist of undigested food and bacteria. It has never actually been inside our bodies. Apart from the bile and one or two other remnants from our exocrine glands, it cannot be regarded as excreta, despite the common use of that word to describe it.

**Bryn Glover**
Cracoe, North Yorkshire, UK


The types of waste we excrete have two different origins. Faeces contains the leftover, indigestible portion of the food we eat, plus bile from the liver, which gives excrement its brown colour.

Urine, on the other hand, is the result of blood filtration in the kidneys. Urine contains nitrogenous waste, primarily in the form of urea, which results from the metabolism of nucleic acids and proteins, separate from digestion. Furthermore, urine also contains water and solutes from the blood, and interstitial fluids – these are excreted to maintain water balance. Essentially, faeces are the result of a coarse, large-scale process of the digestive system alone, whereas urine production occurs at a much finer scale, eliminating wastes produced by all the body's cells.

Our two excretory systems are obvious because we have a separate opening for each: the anus and the urethra. In other organisms, the distinction between the systems isn't as obvious. Animals such as birds, reptiles, and amphibians possess a cloaca, which serves as a common opening for both liquid and solid wastes. Insects blur the line even further: rather than having a distinct urinary system, they rely on Malpighian tubules in their digestive systems – outgrowths of the gut that perform the same filtration function as our kidneys.

**Shaun Hug**
La Mirada, California, US

## ? Smells fine to me

*I recently bought a spray-on deodorant. When I got it home I realised it was intended for women but, not wanting to waste money, I used it anyway. Nothing untoward happened and I received no strange looks from colleagues or friends. So what are the differences between deodorants meant for men and those that are meant for women? How might using the 'correct' deodorant for your sex work better than using one meant for the opposite sex, and what are the pitfalls of applying a deodorant intended for the opposite sex?*

**Alan Ainscroft**
Leeds, West Yorkshire, UK

The primary function of all deodorants is to inhibit growth of bacteria, which feed on secretions from sweat glands. Deodorants are most often differentiated, if at all, by strength. But sex sells, so we have men's and women's. They have only three differences: advertising, packaging and fragrance. Remove fragrance and there are only two differences: advertising and packaging.

Formulations for both sexes contain such fragrances as flowers, herbs, spices, fruits and woods, and are judged as suitable entirely by personal and cultural taste. For example, one upscale new fragrance contains citruses, herbs, ylang-ylang, jasmine and tiare flowers, musk, tropical woods and coconut – and it's for men. And many women buy men's fragrances, because there is none of the social embarrassment that the reverse carries.

**Toshi Knell**
Nowra, New South Wales, Australia

# ? Bags of sleep

*Why do dark circles appear under your eyes when you are tired?*
**Tammy-Ann Sharp**
New Romney, Kent, UK

The key word here is 'tired' – tiredness has much wider connotations than mere sleepiness.

Happy, healthy people who are merely feeling sleepy don't usually show these dark rings beneath their eyes. Sleepy eyes are droopy and somewhat sunken in the sockets. They may also become reddened because sleepiness slows the rate of blinking, causing drier eyes and itchiness, especially if the air is dry and there is cigarette smoke about.

Sleepy people show less facial expression and tend to be poker-faced, but dark rings under the eyes are not present unless these rings were there in the first place, when sleepiness might darken them a little.

Instead, rings are often a sign of being chronically worn out, stressed and run-down – tiredness rather than sleepiness. Tired people don't just need more sleep, but a better and more agreeable lifestyle. This is easier said than done, of course. They may even find that they have difficulty sleeping despite their tiredness, but resorting to sleeping tablets is not the answer.

As for the anatomy of these rings and whether they are due to blood pooling, skin-thinning through dehydration or something else, no one really knows.

**Jim Horne**
Sleep Research Centre
Loughborough University
Leicestershire, UK

## 🛈 Beat generation

*Most music is written in 4/4 metre, giving four beats per bar.
Why are we inclined to prefer 4/4 time? Are there circuits in our
brains that tick along in patterns of four?*

**Michael Light**
Carlisle, Western Australia

*Lots of debate here about whether rhythmic choice is nature or
nurture, as you'll see below. Marching may be the origin of the 4/4
metre, but it is clearly only part of the story, for the reasons outlined
in the final contribution – Ed.*

I think that it's something of an exaggeration to say that most
music is in 4/4, but a great deal of it is, hence its alternative name: 'common time'. As a music teacher, I know many
students find it much harder to play in 3/4 time than in 4/4,
and that they have a strong tendency to insert an extra beat
in 3/4, either by lengthening the last note of each bar or by
leaving a gap after it.

My own feeling, based on years of listening to students
struggling rather than on any knowledge about the brain, is
that we prefer 4/4 because we have two arms and two legs.
March tempo comes naturally to us as an extension of the
movement when we walk and swing our arms. 'Left, right,
left, right' easily becomes '1, 2, 3, 4' and any music in 4/4 can
be thought of as a modified version of march tempo.

I'm convinced that if we had three legs we would find 3/4
time much easier, but that might cause other problems.

**Daphne Gadd**
Cambridge, UK

Popular music, be it either dance or song, does not have to be
in 4/4 time. One need only look at collections of folk music

to see an array of different time signatures. From the Basques to the Bulgarians you can find 5/8 and 7/16 time signatures.

In Latin America, the characteristic sesquialtera rhythm requires juxtaposition of 3/4 and 6/8 time. And let's not start on Indian and African rhythmic complexities, which far exceed anything that can be found in the unsubtleties of western pop and rock music.

The reason 4/4 time became entrenched in popular western music during the 20th century is through the influence of jazz, which owes part of its origins to the marching bands that played at funerals for black people in the southern states of the USA. Most marches are in duple or quadruple time. Since then, the relentless and ubiquitous promotion of modern pop music has perhaps blunted many people's appreciation of other time signatures.

Ironically, in the light of its 4/4 influence elsewhere, jazz retained its diversity with, for example, jazz waltzes. Dave Brubeck's *Take Five* is in 5/4 time and many other virtuosi have used still more exotic time signatures, such as 11/8, or have even expressed two different time signatures simultaneously.

Historically, 3/4 or triple time has probably been more significant. The 19th century favoured the waltz and mazurka while in the 18th century dance music was based on minuets, which are played in triple time, and baroque orchestral overtures that began in slow quadruple time but ended in fast triple time.

The favourite dance of Charles II was in 6/4 time, Louis XIV of France loved the sarabande, Elizabeth I's favourite dance was the galliard, all of which are – like *Greensleeves* – in triple time. Going back to the Middle Ages, church music was written in triple time because it was associated with the Holy Trinity.

So there is no natural predisposition for 4/4 time. It is,

in fact, an illusion forced upon us by pervasive modes of contemporary western culture.

**Ian Gammie**
Corda Music Publications
St Albans, Hertfordshire, UK

If you were living in Vienna in the mid-19th century – the era of the Strauss waltzes – you might well have thought that most music was written in 3/4 time. Both 3/4 and 6/8 times are very common in European classical music, just as they are in classical Indian music.

All evidence suggests that these rhythms are part of cultural inheritance rather than determined by hard-wiring in the human brain. Almost all music derives ultimately from folk song and dance (much of Bach's music is based on dance rhythms) and presumably various rhythms go in and out of fashion. Supporters of the hard-wired brain theory would need to explain the popularity of the often complex time signatures of music from all around the world.

**James Hamilton-Paterson**
Timelkam, Austria

# ❓ Who nose?

*My dad keeps telling me not to pick my nose and eat it. Will eating my bogeys do me any harm?*

**Thomas Walker**
By email, no address supplied

Picking and eating are for cherries. Even if your bogeys do not affect your digestion, chewing them could affect the health of your social relationships. Try chewing gum instead.

Physiologically, eating your dried snot would not matter much. If that solid bogey that you find so toothsome had not dried out, it would have dribbled down your pharynx and been swallowed anyway, unless you intercepted it with your sleeve or handkerchief or stuck it on the underside of your chair.

For the most part, any germs it contained would be digestible or would otherwise die in your gut, but this is not always the case. Some germs do infect people via the nose, and some toxic dust particles do stick in your phlegm. It is to the benefit of your health to ensure you expel such things.

It is not for nothing that your nose hairs stop bugs and dust from landing in your lungs or gut. Blowing your nose would not stop everything, but it is better than guzzling snot.

**Jon Richfield**
Somerset West, South Africa

The medical literature is a delightfully rich source of information about nose picking. Firstly, nose pickers should not feel isolated or guilty about the activity. A US survey in *The Journal of Clinical Psychiatry* in February 1995 of 100 adults in Dane County, Wisconsin, concluded that the activity 'is an almost universal practice in adults'. The same publication in June 2001 carried another survey, this time of adolescents in India, which found that the average frequency of pickage was four times a day.

However, too much of a good thing can prove to be a problem. Rhinotillexomania, or compulsive nose picking, can lead to epistaxis (better known as nosebleeds) or even septal perforation.

The cause of compulsive nose picking is unknown but in extreme situations may transcend habit and become a sign of a psychiatric disorder. The Wisconsin study identified one individual who spent more than 2 hours a day digitally excavating their nasal cavities.

Perhaps more worryingly, Dutch researchers reporting in the August 2006 edition of *Infection Control and Hospital Epidemiology* found that the frequency of nose picking correlated with the presence of nasal *Staphylococcus aureus*, a bacterium carried by about 25 per cent of people but which in its most horrible form can cause lurid 'flesh-eating' infections.

Alas I have not been able to find any studies into the effects of eating one's bogeys, but it is almost certain that there is an Ig Nobel award set aside for anyone who is willing to tackle this important medical conundrum.

**Digbeth D'Marriotti**
By email, no address supplied

## ❓ Dead in space

*During long voyages in space it is possible that people will die, either from illness or because of an accident. What plans are there for disposal of the corpses?*

**Jessica Franklin (age 12)**
London, UK

During long crewed space voyages, such as the journey to Mars that NASA was once proposing to undertake, astronauts will be exposed to numerous risks such as sustained exposure to radiation, as well as the usual health problems people face on Earth.

In light of this, it is perhaps surprising that NASA has no policy for the disposal of the dead on long missions. Interestingly, though, the US National Aeronautics and Space Act 1958 states: 'Each crew member shall provide the Administrator with his or her preferences regarding the treatment accorded to his or her remains and the Administrator shall, to the extent possible, respect those stated preferences.' This suggests that the possibility of death has been considered.

If an astronaut should die, there are three options for the disposal of the body: in space, burial at the destination, or to return the body to Earth.

Burying a body on another planet raises bioethical issues, particularly if there could be an impact on possible alien life. However, travelling on a spacecraft for months or years with corpses might prove uncomfortable for the rest of the crew. Nevertheless, spacecraft designed for long missions may have to include culturally appropriate disposal units or storage areas for the dead.

**Alice Gorman**
Flinders University
Adelaide, South Australia

## ❓ Don't think about it

*As a secretary, my job involves a lot of typing. If I am not concentrating on what I am doing, I can type very quickly and accurately, but as soon as I think about it, where the keys are, for example, I type like a fool, extremely slowly and with numerous errors. The same applies to plenty of other activities, such as playing the piano, driving a car, even reading and talking. If you think about what you are doing you do it less efficiently. Why?*

**Lucy Kaye**
London, UK

This observation is certainly correct and can be seen in controlled laboratory conditions. However, the reasons for it are not clear.

My speculation would be that the difference when doing

something automatically, rather than thinking about it, is similar to the difference between parallel and sequential processing in a computer. With parallel processing we can take account of a variety of things at the same time. But when we think about what we are doing, akin to serial processing, we can only attend to very few aspects of the task. We may even attend to the wrong ones.

However, this leaves us with the mystery of why thinking about what we are doing is often beneficial. If it wasn't, we wouldn't do it. This could be because thinking about what we are doing is the only way to do something novel before the automatic processes have been learned.

**Chris Frith**
Wellcome Trust Centre for Neuroimaging
University College London, UK

# ❓ Cold patch

*Having recently gained weight, I've found that areas of my body where fat has deposited become cold to touch more quickly than those with less fat. Why is this?*

**Katherine Helmetag**
Troy, Michigan, US

To put it simply, fat is an insulator. In the same way that walls insulate a home, fat decreases the rate of thermal loss so its outside edge feels cooler. If you touch a person wearing a winter jacket, you will find it is cooler than their skin.

**M. Sjolander**
Ankarsvik, Sweden

When ambient temperature is lower than that of the body,

the temperature of the skin depends on the balance between the amount of heat reaching the skin, either via warm arterial blood or by direct conduction from underlying tissues, and the amount of heat leaving the skin to the environment.

When we get cold, blood vessels under the skin close up, reducing the delivery of heat to the skin by arterial blood. The skin is then kept warm by heat conduction alone. Fat, or adipose tissue, is a better insulator than lean tissue because it contains less water. Thus any skin above adipose tissue will receive less heat, making it cooler than the skin over lean tissue.

Many animals, especially marine animals, exploit this property of adipose tissue by having a thick layer of subcutaneous fat. When these animals need to conserve heat, the blood vessels below the surface close, as in people, and the thick insulating layer of fat minimises heat conduction to the skin. When heat dissipation is needed, during exercise for example, the flow of arterial blood to the skin is increased, effectively bypassing the subcutaneous insulation.

**Shane Maloney**
School of Biomedical and Chemical Science
University of Western Australia, Perth


# ? I left my heart...

*Why does one really seem to feel the emotional response known colloquially as a 'broken heart' in the middle of one's chest; indeed actually in the region of the heart?*

**Mark Collen**
Sacramento, California, US


A metaphor offers a clue: the Japanese emphasise the stomach rather than the heart. Also, in English we speak of 'not having the stomach' for something. In healthy people it is the muscular viscera that draw attention to themselves, particularly those of the heart, oesophagus and stomach. Though they are under involuntary nervous and hormonal control, their physical reactions give dramatic feedback.

The heart reflects emotions by the intensity and rhythm of its actions, for example in shock it leaps and pounds. Anxiety can cause actual stomach aches and 'lump-in-throat' oesophageal spasms.

Helplessness causes diastolic flaccidity – a drastic reduction in blood pressure – which may explain deaths in those who believe they are the subject of a voodoo curse. It could also explain the physical heartache of grief, loss or betrayal. In addition, it reduces circulation and causes cardiac irregularity or palpitations, with frightening symptoms such as faintness and tingling in the face and extremities.

Conversely, surges of adrenalin cause the pulse to race, which increases blood pressure for emergency exertion but can cause paralysing panic when one does not know what to do.

Compare this to the less muscular vital organs which cannot give such immediate feedback – it takes time to interpret what the liver or kidneys have to say.

**Antony David**
London, UK

# ? Eye level

*The eye views images upside-down in the manner of a camera lens, but our brains reinterpret this input to allow us to see things the correct way up. Have there been any examples of damage to this part of the brain, causing people to see the world upside down? How does this happen, is the brain able to compensate and, if so, how?*

**Kel**
Gladesville, New South Wales, Australia

There is no example of damage to the brain causing people to see the world upside-down. This is because the image itself doesn't actually transfer directly to your brain; only a series of electrical signals is carried there.

The lens of the eye does focus an upside-down image onto the retina. This image is then translated into a series of electrical signals which travel down the optic nerve and pass through the lateral geniculate nucleus – a kind of way station – into the occipital (visual) cortex at the back of the brain.

The reason that the upside-down image does not get flipped is because there is no image to flip. In your brain there are only electrical signals being sent from neuron to neuron, transforming as they go. Your brain processes these signals to create your experience of sight.

Experiments show that if imagery received by the eyes is inverted for a long time, these signals are simply reinterpreted by the brain and eventually perceived as the right way up.

**Gregory Szucs**
North York, Ontario, Canada

The brain does not need any special mechanism to compensate for the image in the eye being upside-down. Once the retina has converted the image into neural information, the physical arrangement of the information is arbitrary.

For example, why should it matter to the brain cells dealing with the top half of the visual world that the nerves supplying them with information happen to originate in the bottom half of the retina?

**Tim McCulloch**
Camperdown, New South Wales, Australia

As a child, I remember constructing a pinhole lens using toilet-paper rolls and tissue paper as the screens. I placed it against one eye and covered my other eye. These lenses invert the image of the world, and initially it was very disorienting seeing everything upside-down: I walked into doors and collided with any number of household objects. However, tactile feedback is a good teacher, and I learned to cope with it. After some time, my brain adapted and the image I was perceiving reverted back to normal. Then, of course, once I had adapted, taking the lens off caused everything to flip upside-down again, until I readjusted once more.

I guess that the ability of the brain to cope with an inverted view of the world would be similar to coping with a mirror-image view: at first, trying to correctly position something while looking in a mirror is very difficult, but with practice it becomes instinctual.

**Simon Iveson**
Warabrook, New South Wales, Australia

It is generally known that our eyes form an inverted image of what we see and that the brain corrects the scene to look the right way up. However, when people wear inverting spectacles so that a scene is inverted before it enters our eyes, the wearer should see the world inverted. George Malcolm Stratton did this experiment in 1897 and claimed that the world looked the right way up again within a week. In other words, the brain 'reinstated' the upright vision.

The experiment has been repeated a few times since, with mixed results, so the jury is still out on this claim. Experiments in the 1940s and 1950s showed that human subjects managed to ride bikes and to go skiing while wearing inverting spectacles, suggesting that they were seeing the world the right way up. However, in the late 1990s a team led by David Linden refuted this claim in *Perception* (vol. 19, p. 469). Their paper suggests that those wearing inverting spectacles simply adapt to seeing the world upside-down by learning new motor patterns and increasing their skill at spatial transformations.

**Mike Follows**
Willenhall, West Midlands, UK

# ? Can't face it

*Why do we grimace when we eat sour or bitter food?*

**Laura Offer**
London, UK

The body has stereotyped sequences of action for avoiding and responding to noxious stimuli such as pungency and acidity. Some resemble involuntary defences against physical attack and are similar through most of the animal kingdom, so they are almost certainly primitive in origin.

In humans, a faceful of ammonia or acetic acid fumes causes retreat, closed eyes and arms thrown across the face, among other responses. A noxious mouthful of a salty, bitter, acidic or otherwise vile chemical that our species instinctively avoids, such as one's own ordure, causes another range of reactions, related to spitting or vomiting. Typical responses include: drawing down the corners of the mouth or gagging in preparation to vomit; salivating to clear the mouth and dilute harmful substances; puckering to avoid more intake; closing the eyes for protection; and performing convulsions that would help free oneself from assault.

More trivial stimulation – for instance, from piquant foods like pickles or mustard – provokes milder incipient reactions such as grimaces and shuddering. Perhaps the reason these different levels of reaction have survived and become entrenched is that such behaviour has evolved into a warning to offspring and associates: 'Bad stuff! Beware!' These less vigorous communication signals evolved more recently than the primitive reactions, and accordingly vary more widely between species, but they serve the same functions: warning of danger or nastiness, or indicating good feeding.

**Jon Richfield**
Somerset West, South Africa

# ? Altered images

*Driving along in the car the other day, my four-year-old son asked why things that were closer to us were moving faster than those further away. What should I tell him?*

**Milton Inverdale**
London, UK

*Thanks for a vast number of answers to this question, many of which were probably more suited to undergraduate level than to a four-year-old. However, one notable group of wags insisted on sidestepping the answer at all costs. Among these was the inevitable 'Ask your mother', from Tony Turner of Tuross Head, New*

*South Wales, Australia. Stephen McIntosh of Hull, UK, suggested: 'You are far too intelligent for a four-year-old... have a lolly.' More encouraging was the answer from Dave Oldham of Northampton, UK, who offered: 'If you can ask a question like that at four years of age it won't be many more years before you can explain it to me.' And congratulations to Peter Gosling of Farnham, Surrey, UK, for his unashamedly literal view of the world. His advice was: 'I think you should tell your son that it is illegal for him to be driving at four years old.' – Ed.*

When my son was a similar age, I tried to explain this phenomenon during a train journey. First I pointed out that objects further away look smaller. I used his hands to show this: if he held one hand close to his face and the other at arm's length, the one at arm's length appeared smaller, even though he could put his hands together to confirm they were the same size.

Secondly, I showed him that it takes more objects to fill the same amount of visual space if they are further away. For example, if the hand further away is half the apparent width of the one closer, it takes two hands to fill the same width.

Finally, I got him to think about something moving, such as an index finger traced slowly from one side of his palm to the other. If it moved at the same speed when it was further away, it travelled the same actual distance (a palm's width), but seemed to have travelled only half as far. So it would take twice as long for it to look like it had travelled the same distance. I then summed up by explaining that the distant things were not actually moving slower, they just looked as if they were.

I also had to explain why it looked as if the trees and houses were moving when my son was sure that they weren't really. First, I got him to move his hand in front of his face, and then to hold his hand still but move his head from side to side. In each case he could see that the hand seemed to move across his vision in the same way. I told him that the two movements were equivalent and he seemed to accept that.

The other passengers on the train thought I was a little strange, but it kept my son quiet.

**Keri Harthoorn**
Stoke-on-Trent, Staffordshire, UK

The answer is that the type of optical system that is used by our eyes causes us to perceive a particular object as 'smaller' the more distant it is – a phenomenon called foreshortening. As our vision system converts the angles subtended by the things we are looking at into apparent distances on our retina, this causes nearby objects to sweep through our field of vision much more rapidly than distant ones. So while distant and nearby objects are within the same field of vision, those further away take longer to pass across it, as they have a low angular velocity, than those that are closer.

You can demonstrate this by placing your hand on a newspaper. Make a 'V' with your index and middle fingers and sweep it along the text. Your hand is the car, and the V is your field of view. You can see that the text near your fingernails takes a long time to move from one finger to the next, while the text closer to your hand moves more rapidly.

**Gregg Favalora**
Arlington, Massachusetts, US

One way to demonstrate this process is to put a toy car on a path representing the road, with an object placed 30 centimetres ahead and 30 centimetres to the side of it. Show your son how the object goes from being diagonally ahead to diagonally behind the toy car when you move it forward 60 centimetres. Then do the same thing, but with the object 3 metres ahead and 3 metres to the side. This time the car has to travel 6 metres to cause the same change in the angle at which someone in the car would view it. Also point out that it takes much longer for the car to travel 6 metres as for it to travel only 60 centimetres.

**Eric Kvaalen**
La Courneuve, France

## ? Think hard

*What is the storage capacity of the human brain in gigabytes? If we were to construct a PC with similar computational power to our brain, what would its technical specifications need to be?*

**John Gladstone**
Southampton, UK

We can only answer this question if we assume the human brain is like a computer. For example, if each neuron holds 1 bit of information then the brain could hold about 4 terabytes (4,000 gigabytes).

However, each neuron might hold more than 1 bit if we consider that information could be held at the level of the synapses through which one neuron connects to another. There are about 50,000 synapses per neuron. On this basis, the storage capacity could be 500 terabytes or more. But these are perhaps misleading answers because the human brain is not like a standard computer. First, it operates in parallel rather than serially. Second, it uses all sorts of data-compression routines. And third, it can create more storage capacity by generating new synapses and even new neurons.

The brain has many limitations, but storage capacity is not one. The problem is getting the stuff in and, even more problematic, getting the stuff out again. We can demonstrate that storage capacity is not the problem if we consider the technique experts use to remember the order of a shuffled pack of cards. This technique, called the 'method of loci', goes back to classical antiquity. It involves imagining a journey in which each card appears at a certain location.

Here is an example that I found on the internet: the first card is an 8 of clubs. To memorise this you imagine going out of your front door, the first step on the journey, and finding your path blocked by a person smashing an egg timer (which is shaped like an 8) to pieces with a mallet (a club). The next card is then placed on the next step of the journey with an equally vivid image.

What is striking about this technique is that the story you create to remember the order of the pack of cards contains much more information than the simple pack of cards you are trying to remember. The vivid images are necessary to get the information into our brain and to get it out again later.

**Chris Frith**
Wellcome Trust Centre for Neuroimaging
University College London, UK

## ? It's his hormones

*What would happen if a man took the contraceptive pill, either once, accidentally, or daily? Are there any published cases?*

**Arnout Jaspers**
Leiden, Netherlands

If he took it only once, probably nothing would happen, except perhaps side effects such as nausea. As for regular, long-term use, the effects would depend on the type of pill he took: combined or progestogen-only.

The combined pill contains oestrogens, which are responsible for the development of female secondary sexual characteristics, so a combined pill would cause slow changes in some body features that would vary from man to man.

The first changes would probably be visible within two to three months. The man's breasts would start to grow, with his nipples becoming larger. His skin would become thinner and softer, which would lead to change in skin tone – pink, with the veins more visible. Body fat would start accumulating in 'female areas', such as under facial skin – making his face look puffier – and also around his hips, thighs, upper arms and pubis.

More significant changes, such as a slimmer waist-to-hips ratio and fleshier hips and buttocks, would probably take many more months to develop. His muscles might become thinner, and his body and scalp hair might change in texture, but the pill alone would not inhibit the growth of facial hair or improve male pattern baldness. Sweat production would also change, as would the body odours of skin, sweat and urine, which would become less sharp and more sweet and musky.

The pill-taker would notice some emotional changes too, such as a greater tendency towards mood swings or depression. Recent studies have indicated that cross-hormone therapy in male-to-female transsexual people may result in a reduction in the volume of the brain towards female proportions – but with no effect on IQ. Regular intake of oestrogens would also increase the risk of blood clotting, decrease insulin sensitivity and cause disturbances in liver function.

By contrast, the hormones in the progestogen-only pill do not cause feminisation in a male. Some studies show that they act like anti-androgens and would probably suppress testosterone to some degree, causing breast growth and a decrease in facial hair. This pill might raise the taker's body temperature and cause fluid retention.

I could not find any clinical studies, but it seems certain that some will have been done, because oral contraceptives have been widely used in gender reassignment for men who want to become women.

**Joanna Jastrzebska**
North Shields, Tyne and Wear, UK

I was a guinea pig for a trial a few years back which involved exactly this, as part of research into a male contraceptive pill. The trial lasted one week, during which some of us were given a common oral contraceptive but at four times the dose prescribed to women, while others were given the same quantity of the same hormones by injection. I was in the second group.

The first day I didn't notice much difference, so I assume a man taking a single pill would be unaffected. The second day, I felt a little down and emotional. My libido started to diminish. On the third day, I was tearful.

I don't know what happened to my testosterone levels, but I wasn't interested in going to the gym any more. More than usual, I wanted to eat chocolate and chat to my female friends. Yes, I am being serious. Little things like events in a movie would start me crying. This stabilised at around the fourth day. On the seventh day I took the last dose of hormone and the effects wore off within a couple more days. The doctor in charge assured me there would be no lasting changes.

All in all, I feel I've had an insight into what it feels like to be a woman. I suspect that it would certainly work as a contraceptive (beyond its chemical effect) because I had no interest in sex, but I doubt many men would want to take female hormones at the expense of their 'manliness'. It was an interesting experience, but not one I wish to repeat.

**R. Ross**
Aberdeen, UK

# ? Taste tribulation

*Readers have often contacted us to ask whether there is anything Jon Richfield – a prolific answerer of questions from these books – does not know. Well, it seems there is – Ed.*

*This is a question that my husband, Jon Richfield, cannot answer to my satisfaction. I find the taste of certain common spices quite horrible. The nasty flavour I get from all of them seems, to me, quite similar. The spices that taste this way are aniseed, caraway, cumin, fennel and coriander. Tarragon, cardamom and capers also taste awful in the same way. I wonder if there is a food scientist who knows what they have in common, or what my aversion might be. I should add that I am not a fussy eater in general.*

**Bess Richfield**
Somerset West, South Africa

The substance that your correspondent finds foul-tasting is probably a terpenoid called carvone. This chemical provides the principal flavour of caraway seed and, by itself, has a typical caraway flavour.

All the spices she mentions are members of the Umbelliferae family, and all of them contain this substance. Carvone is also found in tarragon, cardamom and capers.

**Ralph Hancock**
London, UK

I heard once, but have not verified myself, that there is a cis-trans isomerism (the orientation of groups of atoms and the positions they occupy within molecules) in the taste buds of some people which binds to at least one of the flavour components of cilantro, the leaf part of coriander.

For most people, the taste is pleasant. For those with the other isomeric version of the taste bud's active site, however, the taste is more like soap. If this theory is correct it could extend to the flavour experience of the other spices such as aniseed and caraway.

**Karen Nyhus**
San Francisco, California, US

One explanation for the perverted taste sensations with some spices is that your correspondent suffers from mild zinc deficiency. This gives rise to hypogeusia, or altered taste, usually for the worse. It would be interesting to find out if taking zinc sulphate tablets, under the direction of a doctor, for four weeks would make a difference.

**Tim Healey**
Barnsley, South Yorkshire, UK

## ? Pint of the usual?

*The first time I had two pints of beer in my late teens I was horribly sick. Now I can drink two pints of beer without feeling any ill effects. What is the mechanism by which our bodies become tolerant to alcohol, or indeed other drugs, all of which have a smaller and smaller effect with regular use? After all, I am consuming exactly the same amount of poison which made me ill 30 years ago – why doesn't my body just do what it did back then?*

**Rob Howe**
Gomersal, West Yorkshire, UK

The ethanol in alcoholic drinks is metabolised almost exclusively by the liver. The liver enzymes responsible are alcohol dehydrogenase (ADH), catalase and an enzyme complex known as the microsomal ethanol-oxidising system (MEOS).

Repeated administration of alcohol will increase the amount of these liver enzymes, and subsequently improve its ability to metabolise alcohol.

However, such enzyme induction does not fully account for improved alcohol tolerance. There is also a mechanism of behavioural tolerance whereby an individual learns to function under the influence of alcohol.

Finally, you are probably larger now than when you were in your late teens. This means that your total blood volume is also likely to be increased, so although two pints of beer will contain the same mass of alcohol, it will be at a lower concentration in your blood.

**Benjamin Hunt**
Leicester, UK

One explanation for the development of tolerance stems from a 2004 study at the University of Texas into fruit flies' response to benzyl alcohol. As the fruit flies developed tolerance to the alcohol, researchers noted an increased activity in their *slo* gene. *Slo* modulates a cell-surface protein, helping to increase signalling between nerve cells in the brain. Under sedation from alcohol, the activity of the fruit flies' *slo* gene doubled.

The protein acts almost as a thermostat would. For example, if a drug other than alcohol excites the nervous system and increases signalling, *slo* gene activity decreases, which suppresses the effects of the drug. Conversely, alcohol suppresses the nervous system and *slo* activity increases, serving to counteract the sedative effects of alcohol by stimulating signalling.

Previous exposure to a drug enhanced the future performance of *slo* in the fruit flies, meaning it becomes more responsive and better able to suppress the effects of a drug.

The *slo* gene found in fruit flies is very similar to the human version. The performance of the questioner's genetic thermostat will have improved after 30 years of practice, and therefore more pints are required.

**Jeremy Fancher**
University City, Missouri, US

## ❓ Up in smoke

*How many people are cremated each year and how much energy is consumed in the process? Will these numbers increase on current projections? And are there no better and more environmentally friendly methods of disposal?*

**Jeremy Dawson**
Laurencekirk, Grampian, UK

Cremation is the accepted practice for disposing of dead bodies for a number of religious groups in India, including Hindus, Sikhs, Buddhists and Jains. Some 85 per cent of the country's 1-billion-plus population cremate their dead. Beyond the Indian subcontinent, Hinduism is also a significant religion in Mauritius, Guyana and Fiji.

Many people from other religions, including Taoism and Shinto practised in China, Japan and Indo-China, also choose to cremate their dead. Cremation is not just limited to religious custom; it is also practised by many people in Europe and America for reasons other than religion.

A conservative estimate for the number of people around the world who would opt for cremation is around 1.2 billion. Taking an annual death rate of 1.5 per cent, that means roughly 18 million cremations annually.

It takes about 100 kilograms of wood to create a fire that is hot enough to cremate an average human body, so that adds up to 1.8 million tonnes of wood. If we take the energy value of wood as 17 megajoules per kilogram, this works out at about 30 million gigajoules.

In some places, electric furnaces replace wood. These tend to have a capacity of between 75 and 100 kilowatts, and they can cremate a body in 30 minutes or so, consuming somewhere between 0.13 and 0.18 megajoules for each body.

However, most bodies worldwide are simply buried. Coffins consume wood, of course, and the quantity can be substantial in prosperous western societies.

There are many factors to consider regarding the environmental impact of these various practices. While cremation consumes wood and generates smoke and carbon dioxide, burial also generates carbon dioxide, methane and other substances as bodies decompose, although this is spread over a longer time period.

In addition, burial also occupies land, and in some societies such land is considered sacred and so cannot be used again. At one time, families in China reserved the most fertile patch of their farmland for burials, thus blocking land use forever.

Apart from burial and cremation, there are some unusual practices in some parts of the world. Some Tibetans hack up bodies and feed the pieces to vultures. In India and Iran, Zoroastrians also allow their dead to be consumed by vultures, placing the bodies on a circular structure called a tower of silence. Tibetans and Zoroastrians believe that the body should serve a useful purpose after death – and the environmental impact is zero. Burial at sea, as practised by sailors, also works in the same way, although the motives are more practical.

Some people in other societies have the same thought regarding usefulness when they donate their bodies to medical research. However, the remains have to be disposed of later, normally by incineration.

**Dileep Paranjpye**
Lucknow, India

The most recently available global figures for cremation are from 2006 and they put the figure at 7,838,353. As with most global statistics, the heavyweights are China, which cremates more than 4 million people a year. In places with little spare land, such as Japan, nearly everyone is cremated, while in the US only 33 per cent of people are. These figures are likely to increase as the Catholic population becomes more comfortable with cremation.

Calculating the amount of energy used in cremations worldwide is nearly impossible because of differences in fuel types and costings.

Cremation is, of course, becoming untenable. The gas used is a fossil fuel creating heat and pollution, and the vapours emitted, despite increasingly sophisticated and expensive filters, release toxins such as mercury.

Susanne Wiigh Masak, a Swedish biologist and keen gardener, has invented a process she calls Promessa, in which liquid nitrogen is used to freeze-dry bodies into what she describes as a perfect compost.

Another alternative, marketed by Sandy Sullivan, is called Resomation. This is a form of speeded-up anaerobic digestion using heated alkalines. Both are hugely improved methods of disposal, which have interesting possible environmental applications. The major obstacle is, of course, the squeamishness of the public.

The most environmentally sound method of cremation is on an open-air pyre using wood. Not only would this be carbon neutral, but it would be much more spiritually and psychologically nourishing than the current industrial conveyor-belt approach that is used in most modern crematoriums. It is my company's most-requested 'fantasy funeral'.

**Rupert Callender**
The Green Funeral Company
Dartington Hall Estate, Devon, UK

# ❓ Wotsisface?

*Why, after I've spent hours attempting to remember somebody's name or something similar, does the answer eventually arrive in the middle of the night when I'm not even trying?*

**Ben Longstaff**
London, UK

It has been suggested that when someone has this kind of sudden insight (an 'aha!' moment), one's mind has taken unconscious 'pathways' that have led to the solution of the problem – whether it's your cousin's boyfriend's name or 5-down in the crossword you attempted yesterday.

It seems that the first time you were trying to remember that name, however, your mind activated the wrong pathway. That misdirected activation might have been stronger than the answer-related activation, masking the latter, even though you knew the answer. Only when the former subsides can the solution-related activation surpass the threshold of consciousness and be perceived. It might happen when you're not expecting it, like just before sleeping. More information on these processes can be found in a *Psyconomic Bulletin and Review* article published in 2003 (vol. 10, p. 730).

**Dilza Campos**
Rio de Janeiro, Brazil

Just because the conscious mind is not focused on recalling a name does not mean the brain is not churning away at the problem, even during sleep. Indeed, as a designer I have learned to trust this non-conscious, problem-solving process. Upon retiring, I will often select some difficult, unsolved design dilemma from a current project and 'assign' it to myself. When I awake in the morning, almost invariably I will discover that I have worked out a solution.

Many older people – myself among them – whose memories may be increasingly cluttered and whose recall mechanisms may be slower, discover that precisely by not trying to recall a name or term but merely waiting or continuing with another thought or activity, the sought-after memory comes to them of its own accord.

**Larry Constantine**
Department of Mathematics and Engineering
University of Madeira
Funchal, Portugal

## ❓ Studying form

*When closely matched athletes are competing in events that involve running, swimming, throwing or lifting, why does one of them win one day and another the next? Surely whoever is the fastest or strongest will remain so, for a while at least. Often the original winner will return a few days later and win again, so why did he or she lose the race between the two victories?*

**Magda Loncic**
Kiev, Ukraine

The questioner must not belong to a gym. If you regularly take exercise in a quantifiable way, you soon notice that there are 'strong' days and 'weak' days. The most obvious factors influencing these are the amount and quality of sleep achieved beforehand, and how one is nourished. Elite athletes may eliminate variation in their food intake on race days, and may even manage to regularise their sleep schedules, but one or another may be fighting off a minor viral infection, or the temperature or humidity may be more to one's liking than to another's.

**Steve Gisselbrecht**
Boston, Massachusetts, US

The answer to this question lies in another question within the query: how closely matched are the athletes?

If the athletes were identical in skill, strength, speed and motivation, presumably they would never beat one another. But they do, because there are always factors both innate and environmental that interfere.

Nowadays most athletic events that involve covering a distance are timed to one-thousandth of a second, so any competitor who beats another by less than this margin is declared to have equalled the other's time. That is extremely unlikely, so in fact there is almost always a winner and a loser.

At a more macro level, the environment, plus the athletes' mental and physical states, are likely to vary by more than, say, 0.1 per cent and this makes all the difference at the elite level. When you add in diet, will to win, fit of shoes, distance from starting gun and such like, it's surprising there are any dead heats at all.

**Mike Rennie**
Professor of Clinical Physiology
University of Nottingham Medical School
Derby, UK

An athlete's form reflects many different and constantly varying factors, any of which could prevent a win if overdone, underdone, or done in the wrong combination.

Whether psychological or physiological, any changes in the body's status will take time to resolve, causing cyclical or quasi-cyclical performance as they do. Effective coaches try to time the optimum for the day of performance and the optima may be brief, sometimes ended by the reaction to victory itself – for example, through overconfidence or a lapse in commitment.

Athletes don't perform according to fixed standards of precision like machine tools. The bell curve, or normal distribution, is ubiquitous. It describes the statistical variability of any athlete's performance and of differences between athletes. Each component variable has its own normal curve, and how they combine affects the athlete's overall variability. Different athletes' curves of variation can overlap far enough to reverse the outcomes of competitions dramatically.

Some variation in form is unpredictable, such as injury, illness, personal events, psyching or luck on the day. Other effects are systematic, such as maturing, declining or cumulative stress, whether mental or physical. Any of those could cause gradual eclipse; almost any could cause sudden loss of form, even mid-game, and most setbacks are at least temporarily reversible.

**Jon Richfield**
Somerset West, South Africa


## ❓ Fish suppers

*How do traditional Inuit avoid scurvy?*

**Thom Osborne**
Granada, Spain


Humans – along with other primates, guinea pigs and fruit bats – cannot produce their own vitamin C and so need to get at least 10 milligrams per day from their diet to stay healthy. A deficiency results in scurvy, but it can take several weeks or months before the body shows signs of the disease – starting with bleeding gums and progressing to death if left untreated.

The Inuit avoid scurvy as they, too, get all the vitamin C they need from their diet, especially from eating raw meat. Muktuk – a mixture of frozen whale skin and blubber – is the richest source: 100 grams of muktuk yields 36 milligrams of vitamin C. Weight for weight, this is as good as orange juice. Raw caribou, kelp and more whale skin also provide more than enough vitamin C. The Inuit practice of freezing any food that is not eaten raw helps to conserve vitamins, in contrast to cooking food, which destroys vitamins.

**Mike Follows**
Willenhall, West Midlands, UK

## ？ Swallow your pride

*I've just seen a sword-swallowing act. Swords that were seemingly longer than the depth from throat to anus were swallowed. It has to be a trick, doesn't it? If it is, what's the trick? If it isn't, what's going on?*

**Alan Finnegan**
Whitehaven, Cumbria, UK

The taller you are the more sword you can swallow, but it cannot go past the pit of the stomach. And that is one thing that sword swallowers have to get used to – the feeling as the point of the sword arrives there… just touching, and no more. This is why some swallowers eat heavy food just before performing, to stretch the stomach a bit so they can swallow a longer sword.

The other thing they have to get used to is the gagging reflex when they start to swallow. This can be controlled eventually.

Swallowers use silk to clean the sword just before it's inserted to remove any dust, and again while the sword is being withdrawn (with a great flourish of silk) to clean off acid stomach juices that can attack the steel. It's all a matter of skill and nerve, with little room for trickery.

For the best account of this and more, including swallowing giant corkscrews that make the pharynx jump up and down as they twist; how to swallow neon tubes to make the chest glow from inside; and how to eat and swallow fire (when learning, have a lot of ice cream handy), can be found in *Memoirs of a Sword Swallower*, by Dan Mannix, first published in 1951, which is when I bought my copy. I'll never forget that opening sentence: 'I probably never would have become America's leading fire-eater if Flamo the Great hadn't happened to explode that night in front of Krinko's Great Combined Carnival Side Shows.'

**Michael Boddy**
Binalong, New South Wales, Australia

There is no trick to a genuine sword-swallowing act, or rather not the kind of trick your correspondent is probably thinking of. Strictly speaking, a true sword swallower doesn't actually swallow the sword, but that apparent contradiction is the key to how the swallower actually manages to get the blade all the way down.

To join the Sword Swallower's Association International, a would-be member has to demonstrate the ability to 'swallow' a non-retractable, solid steel blade that is at least 2 centimetres wide and 38 cm long. With those qualifications, it's not really surprising that the current worldwide membership of the SSAI is restricted to a few dozen full-time professionals and a handful of amateurs. Nor is it a surprise that, despite the SSAI's strict membership criteria, many people believe a trick blade is employed, particularly when one learns that the record length for a swallowed blade is an eye-watering 82.5 cm.

The performer, through a regime of practice, learns to suppress the natural gag instinct, by relaxing the upper oesophageal sphincter – which normally closes the throat to prevent us choking or drowning. To do this, they usually start by forcing their fingers down their throat and then work their way through a range of longer and bulkier everyday objects, before regularly exercising with a carefully folded wire coat hanger.

Curiously, in sword swallowing, as in so many other aspects of life, size apparently doesn't matter. Although the artiste who swallowed the aforementioned 82.5-cm sword was 220 cm tall, it is the configuration of a performer's insides that determines the length of blade they can swallow. Particularly critical is the angle of the gastro-oesophageal junction, or cardia – the point where the oesophagus joins the stomach.

The cardia is where the lower oesophageal sphincter is found. This sphincter prevents gastric juices flowing up out of the stomach into the throat. It is the ability to exercise control over this valve, and keep it relaxed when it should, by reflex, close, that is critical for a performer.

It is scarcely surprising that industrial injuries among sword swallowers produce a distinctive pathology. While no member of the SSAI has died as a result of a performance going awry, at least one sword swallower has brushed the side of his heart with a blade, and perforations and lacerations of both the oesophagus and pharynx are common, as are lower chest pains. These pains are often associated with a dramatic technique known as 'the drop', where the sword is downed in one smooth action, controlled only by the muscles of the pharynx.

The most widespread occupational hazard suffered by sword swallowers is a sore throat – 'sword throat', as it is known. At least one swallower had to terminate their career through losing the ability to salivate, as saliva is the principal lubricant employed to ease the passage of the blade (although butter has been used as a substitute).

Of course, this is an experiment that would-be researchers should never, ever attempt at home.

**Hadrian Jeffs**
Norwich, Norfolk, UK

*In February 2010 Australian performance artist Chayne Hultgren, also known by his stage name The Space Cowboy, set a new world record by simultaneously swallowing 18 swords, each nearly 75 centimetres long. 'Wow, I did it, it feels good, thank you very much, it feels really good actually,' he said after setting the record. We still suggest readers don't follow his example – Ed.*

## ? Heated argument

*Why is it that when I get into a bath at 39 °C I feel totally relaxed and yet, when I enter a room at the same temperature, I feel totally stressed?*

**Alan Parr**
Corwen, Clwyd, UK

Although the bath may be at 39 °C, the air in the bathroom is probably much colder, allowing part of the body to lose excess heat. This heat loss is helped by evaporation of the bath water. In contrast, when the air in a room is at 39 °C, heat will be flowing into one's body, rather than away, and body temperature will begin to increase. Since the body then has to lose excess heat, the feeling of stress goads the brain into taking action, such as drinking a cold drink or moving to a cooler room.

**Keith Lawrence**
Staines, Middlesex, UK

# ▆ Oooer

*Why do people say 'um' and 'er' when hesitating in their speech?*

**Michael Larcombe**
Belper, Derbyshire, UK

This question can actually be split into two: why do people say anything at all while hesitating and why do they say 'er' and 'um' instead of other possible sounds?

To answer the first question, linguists known as conversation analysts have observed that people vocalise in a conversation when they think it is their turn to talk, and there are several ways of negotiating the taking of those turns. One of them is the relinquishing of a turn by the current speaker and another speaker taking the floor. Therefore, silence is often construed as a signal that the current speaker is ready to give up his or her turn.

So, if we wish to continue our speaking turn, we often need to fill the silences with a sound to show that we intend to carry on speaking. If we always thought out thoroughly everything we were going to say in a conversation, or memorised our lines perfectly, there would be no hesitation at all.

But, as it is, we do a lot of what is called local management, or improvisation, during conversation for many reasons – not least because we cannot predict the reactions of our interlocutor. In order to keep the floor while we hesitate, we place dummy words in the empty spaces between our words, much as we might drape our coats on a seat at the cinema to prevent others from taking it.

The second question, as to why 'er' and 'um' are used instead of, say, 'ee' or 'choo', is not as easy to answer. 'Er' in British English is a transcription of the phonetic 'schwa' sound found in unstressed syllables of English words (such as the vowel sound in the first syllable of 'potato').

In traditional phonetics this was called the neutral sound because it is the vowel sound produced when the mouth is not in gear, that is, not tensed to say any of the other formed vowels such as 'e'.

The 'um' sound is more difficult to explain unless it is just a bad transcription of the same neutral sound with a consonant that closes the mouth in preparation for another real word.

By the way, these sounds are not universal. Many speakers of other languages hesitate in other ways. In Latin languages, for example, the pure sound of the vowel 'e' is often used.

**William DeFelice**
Barcelona, Spain

'Er' is used as a conversation filler because it is the most easily pronounced voiced sound for an Anglophone. This is shown easily in a stress-timed language such as English in which all unstressed vowel sounds tend towards the central vowel position.

It is known as the central vowel position because it is pronounced in the centre of the mouth, irrespective of the written vowel, as in 'America', 'trousers', 'ferocious, prospective', 'purpose'. 'Um' is really only 'er' with a closed mouth, as can be shown empirically.

English-speaking pupils learning foreign languages have a tendency to 'um' and 'er' in a way which is quite foreign to native speakers of the target language. It is also the case that using the correct alternatives gives an impression of fluency greater than that shown by pupils who avoid such utterances, but whose pronunciation is almost flawless.

**John Gillespie**
Modern Languages Department
Glenalmond College
Perth, UK

'Um' and 'er' are culturally determined. For example, Mandarin Chinese speakers often say 'zhege zhege zhege' (this this this). Some young, hip foreigners learning Mandarin soon 'zhege zhege zhege' with the best of them.

**Kevin McCready**
Wamboin, New South Wales, Australia

People don't say 'um' and 'er' any more. Instead, they say 'basically …'

**R. J. Isaacs**
Barnet, Hertfordshire, UK

Even 'basically' has been superseded by the latest teen speak. It's not 'um' or 'er', it's 'I was like, omigod, that's like so totally not good…'

Fluttering your hands in front of your face as if to cool yourself down at this point is, like, a totally optional extra.

**Natalie Henwick**
London, UK

## ❓ Pointless?

*Ever since the recent birth of my son, who I breastfeed, I have wondered why men have nipples.*

**Dorthe Mindorf**
Aalborg, Denmark

Many suggestions for this phenomenon have been offered. They may exist to help men check that their vests are on straight, or be present as a safety feature – to warn us how far out from the beach we can safely wade.

However, there is a more plausible explanation. Male

and female human embryos are identical in the early stages of their development. If the fetus receives a Y chromosome from its father, a hormonal signal is produced: the labia fuse to form a scrotum, the gonads develop as testicles and a male results. Otherwise the 'default' female remains.

Various structures in the adult reflect the symmetry of male and female and their common embryonic source. Men have nipples because they have already begun to develop when the 'switch to male' signal is received. The development of breasts is halted in most – but not all – cases but the nipples are not reabsorbed.

Another effect of these developmental pathways which are shared by both males and females is pointed out in Stephen Jay Gould's essay 'Male Nipples and Clitoral Ripples' (which can be found in the Penguin Books' 60th anniversary collection, *Adam's Navel*). Males need plenty of blood vessels and nerve endings in their penises to achieve erections. Because the penis and clitoris have their origins in the same structure, females have the same number of blood vessels and nerve endings packed into a much smaller area, resulting in the enhanced sensitivity of the clitoris.

Conclusive evidence that God is not a man?

**Jim Endersby**
University of New South Wales
Sydney, Australia

# ? Ringing home

*What is it about loud rock music that makes your ears ring after several hours at a concert or club? Does loud classical music have the same effect?*

**Rebecca Wiseman**
London, UK

... er, pardon?

**Dave Walters**
Broadcast Engineer
Classic FM, London, UK

Ringing in the ears is called tinnitus. In the circumstances described above it can often be accompanied by a temporary loss of hearing known as temporary threshold shift and indicates damage to the nerves of the inner ear caused by excessive noise exposure.

The criterion is not the type of music listened to but the sound energy involved. Rock music, being prone to electric amplification, can generate high noise levels. This tends not to be the case with classical music, with the possible exception of Wagner.

Repeated exposure to high noise levels can lead to permanent and irreversible hearing damage and because of this, noise in the workplace is subject to legislation – the maximum permitted exposure level being 90 dB(A) over an eight-hour shift.

Damage of this nature tends to affect hearing mainly at a frequency of about 4 kilohertz and presents a characteristic profile on an audiogram.

This loss is insidious and may not be noticeable to the sufferer until natural hearing loss through ageing (presby-cusis) begins to occur. At this point, hearing function may start to drop off sharply and it is, unfortunately, far too late to do anything about it.

It is best not to expose one's hearing to very high noise levels, but if it does occur, it is important to allow full recovery (16 hours) before re-exposure to the sound. You may wonder if rock musicians suffer from this form of deafness – they do.

**Tom Croskery**
Coleraine, Londonderry, UK


## ❓ What's your poison?

*I know people who insist that certain types of alcoholic drinks put them in specific moods when drunk – such as emotional, violent or confident. Is there any scientific reason why different beverages would have specific effects on mood?*

**Frederick Allen**
Oxford, UK

Unfortunately, there is no straightforward evidence to support this claim, nor is there any evidence against it. Whether you're drinking wine, beer or spirits, the alcohol in your drink will be ethanol, which affects several neurotransmitters involved in determining mood. For example, alcohol inhibits glutamate receptors, which has the effect of relaxing muscles; it stimulates receptors that respond to gamma-aminobutyric acid (GABA), reducing anxiety; and it increases the release of dopamine, a hormone associated with excitement.

Mood and behaviour depend also on the degree of intoxication, which can be quantified by measuring the volume of alcohol in a given volume of blood, better known as the blood alcohol concentration (BAC). BAC depends not just on the amount of alcohol ingested but also on gender, weight and body fat.

When BAC is low (up to 0.06 per cent), the effects usually manifest themselves as euphoria, talkativeness and increased self-confidence. With BAC between 0.06 and 0.2 per cent, you will experience excitement and disinhibition, and then mood swings, particularly involving anger, boisterousness or sadness. The next stage, with BAC over 0.21 per cent, brings general inertia and a reduced response to stimuli. If you carry on drinking you will end up in a coma (BAC above 0.35 per cent) or even in the mortuary (above 0.50 per cent).

The context in which alcohol is consumed also plays a role. We tend to drink particular alcoholic beverages in particular situations: fine wine is usually savoured over a nice meal, for example, and hence is likely to put you in a mellow mood, while numerous shots of vodka may be consumed at a party on an empty stomach and will make you feel drunk much quicker.

Some people suggest that the mood you end up in when you drink depends on the mood you are in when you start, and that people tend to choose specific drinks for specific moods.

**Joanna Jastrzebska**
North Shields, Tyne and Wear, UK

## ❓ Ear wiggling

*I am fortunate enough to be able to wiggle my ears. However, I can only wiggle both at once, not one at a time. Why?*

**Peter Slessenger**
Reading, Berkshire, UK

Bilateral symmetry is the default mode for movement. Infants suck, cry and wave their arms symmetrically and must eventually learn to do things one-sided. I have heard youngsters complain that they can't wink: when they try, they close both eyes. Even as adults, it is easier to do mirror-writing with your left hand if you simultaneously write the same word with your right. I, too, could once wiggle my ears only both at once. With practice I learned to wiggle one at a time, an accomplishment of no value to anyone – until now.

**Spencer Weart**
Hastings-on-Hudson, New York, US

Some combined bodily actions share neural channels, which prevent independent action. It is hard, for example, to direct your eyes independently. Physically it should be possible, but your mental control specialises in binocular coordination.

As a rule, independent direction of sensory organs is suited to detecting prey or danger, while symmetrical sensing permits precise measurement.

Most primates use their ears to supplement binocular vision or for direction finding. This means that not many need to move their ears much and hardly any need to move them independently; instead they move their heads.

Correspondingly, visual ear signals such as twitching, vital to most carnivores and many herbivores, hardly figure in the social behaviour of primates, especially the anthropoids. Our legacy is generally symmetrical.

**Jon Richfield**
Somerset West, South Africa

In order to wiggle one ear at a time, practice is needed in front of a mirror. That is how I learned the art. By grinning forcefully and widely, the ears are made to move. If you concentrate on finding the muscles that move the ears, you can operate them without grimacing. Then practise moving each ear by itself. What use does this skill have? It impressed teenage girls, up to a point, and a by-product was the smoothing out of wrinkles on my forehead.

**Brian Colless**
Palmerston North, New Zealand

##  Wakey, wakey

*Why is it that when we are tired the blood vessels in our eyes are more visible?*

**Lucy Bennett**
By email, no address supplied

Apart from causing droopy eyelids, sleepiness slows down blinking, a process which normally keeps the conjunctiva – the outer layer of the eye – moist and well lubricated with fluid from the tear ducts. Its drying out triggers mild inflammation. The more obvious effect is red eyes, a consequence of the dilation of the conjunctiva's capillary blood vessels, which are usually invisible.

All this causes the eyes to become itchy, and rubbing them only makes things worse, as does a dry indoor atmosphere or smoke. Contact lenses become unbearable by this stage, and if they dry out, too, can cause painful scratching of the conjunctiva.

Other than trying to remember to blink more frequently or going to bed and having a good night's sleep, going outdoors into cooler and moister air helps, as does the humid air from a warm shower (though remove contact lenses first).

The quick fix of resorting to eye drops to reduce the inflammation and then going straight back into a dry, smoky atmosphere would be a short-sighted approach (pun, of course, intended).

**Jim Horne**
Sleep Research Centre
Loughborough University
Leicestershire, UK

# ? In a spin

*Why don't adults enjoy dizziness like children do? When I was a kid, I remember thinking that adults were rather boring for not enjoying the feeling of dizziness like I did, and I vowed to always enjoy it. Now, as an adult, I can't stand it – it makes me want to throw up. It seems many other adults feel the same way. Why is this? Does something change in us as we age?*

**Ivan V.**
Mexico City, Mexico

I still remember my first – and so far last – trip to a fairground. I was 15 and vomited after a ride on a merry-go-round. I couldn't understand why my brother, who is three years younger than me, stayed for another ride.

Children obviously enjoy the feeling of dizziness – just

look at how roundabouts in parks and playgrounds are packed with youngsters. They need that stimulation to develop a healthy balance system, which is necessary to crawl, walk and keep their bodies upright, even on a rocking boat.

Our balance system is controlled by three senses cooperating in complex harmony. The vestibular system in our inner ear informs us about the position of our head; our eyes tell us how our body is located in relation to the external world; and proprioceptors – receptors in muscles and joints – help us to figure out how our body is positioned in space, which is particularly helpful if we cannot see. These elements mature at different rates.

The vestibular system is fully operational by the time a child has reached 6 months of age; proprioceptors need three or four years more. The development of the visual element is complete by around 16 years of age.

The sensation of dizziness and nausea following a spinning movement is similar to motion sickness – a result of the conflicting information our brain receives from the three elements mentioned above.

When our body is rotating at speed our vestibular system and proprioceptors can feel it, but our eyes can't locate the horizon. Our brain is desperately trying to resolve this conflict and, because humans are primarily visual, it assumes that the other senses are hallucinating, probably because of intoxication. So the brain tries to get rid of the assumed poison by provoking vomiting.

It looks as if my brother's balance system hadn't fully matured at the time of our trip to the fairground, hence his brain wasn't perceiving the sensory information as conflicting. Therefore, he could enjoy his ride on the merry-go-round while, unfortunately, his older sister could not.

**Joanna Jastrzebska**
North Shields, Tyne & Wear, UK

## ❓ Wine on the line

*Since my 20s, I have drunk on average a bottle of wine a day. I'm 57. That's 49 UK alcohol units a week. The UK's recommended weekly limit for a man is 28 units. I recently had a health check at my local clinic, and I'm in perfect health. Specifically, my liver function tests are entirely normal. Am I exceptional or are the government limits spurious? I rarely drink spirits and occasionally substitute beer for wine. I play football and squash. I walk 3 kilometres to and from work. I lead a normal life and, probably due to regular consumption, I never feel drunk, but presumably I am considered a binge drinker. I don't want advice from a government minister or associated medic. I want objective information. Am I lucky? Or is my consumption relatively harmless? What's the truth?*

**David Hunte**
London, UK

*A UK unit is 10 millilitres (8 grams) of alcohol – Ed.*

The questioner may not be getting away with his alcohol consumption as lightly as he thinks. The liver has a remarkable ability to carry on working, and liver function tests may remain normal even when the organ is quite badly damaged. The gamma GT test is more sensitive than other enzyme tests at detecting damage, but it is often not offered to National Health Service patients in the UK because of its cost.

I also think the writer has underestimated his consumption. If he really drinks a bottle of wine a day then, given the strength of typical popular wines, I would estimate that he could be drinking more than 60 units a week. In more than 30 years of general practice almost everyone I encountered drinking more than 40 units a week was damaging his or her health in some way, through addiction, hypertension, liver or gastric problems, or mental problems.

That said, government advice on alcohol consumption is necessarily arbitrary, and there is great genetic variation in the way that people metabolise and tolerate alcohol.

**Anthony Daniel**
General Practitioner
Sevenoaks, Kent, UK

The short answer is that, like a 90-year-old smoker, you are just lucky. The government limits of 2 or 3 units per day for women and 3 or 4 units per day for men are based on epidemiological evidence. The complex mix of factors influencing our health makes it impossible to issue cast-iron predictions of what will happen to a particular individual at a given level of consumption.

One bottle of wine typically contains 10 UK units or about 80 grams of alcohol. Drinking a bottle a day has been shown to increase the risk of liver cirrhosis 18-fold. You are also five times as likely to get cancer of the oral cavity, and two to three times as likely to get laryngeal or oesophageal cancer, to have a stroke, or to suffer from essential hypertension or chronic pancreatitis.

These are relative risks. What they are relative to will depend on a number of factors, including genetics. The liver is vulnerable to excess fat, so an active lifestyle and low-fat diet will reduce the risk of liver disease.

**Rachel Seabrook**
Institute of Alcohol Studies
St Ives, Cambridgeshire, UK

# 3 Domestic science

## ⁇ Rip off

*Try tearing a piece of sticky tape across its width and you're asking for trouble but, if you nick it with a sharp object first, it tears easily. Why?*

**Daniel Albert**
Science Museum, London, UK

When a material has a crack in it, any stress concentrates around the crack tip. The same applies to sticky tape. The sharper the crack the more stress becomes concentrated at the tip. Hence, even under small loads a crack can propagate through the material. Drilling holes around crack tips in metals or plastics stops them propagating because the crack tip blunts and is less liable to spread.

**Alex McDowell**
South Ruislip, Middlesex, UK

Imagine stress as lines of tension within the stretched material: those lines must detour round the ends of any crack they meet. The deeper and sharper the crack, the more crowded the stress lines passing its sharp edge, so stress concentrates there. In rigid material the concentrated stress favours crack growth. Conversely, yielding materials deform cracks, blunting them, dispersing stress and stopping their growth.

Modern plastic tapes of the type described are made of

polymers whose long molecules have very little give to them when they are stretched. Manufacturers stretch the tape, orienting the molecules to lie straight and closely parallel in the long direction of the tape. Unoriented tape is weak, but tends to distort and stretch rather than snap. Oriented tape is strong until it is nicked. At the nick its oriented molecules resist distortion and the crack propagates across it as though through scratched glass, only more slowly.

**Jon Richfield**
Somerset West, South Africa

This is a simple demonstration of stress concentration. The nick causes a reduction in cohesive surface area, allowing the force of the tearing to act along the geometric discontinuity. You will notice that a piece of tape with a hole created by a hole punch is harder to tear than one with a slight cut along the edge – the round hole distributes the force evenly.

Stress concentrations need to be accounted for by engineers during design. Many have learned the hard way. Nineteen US-built Liberty cargo ships used to supply the UK during the Second World War cracked in half, with the splits originating at the square corners of hatches. The design of its successor, the Victory ship, used circular portholes and rounded hatchways. The de Havilland Comet aircraft also suffered accidents caused by stress concentration. In 1954, one broke apart and crashed en route to Cairo. The cause was cracking around the rivets at the corners of the square windows. Aircraft now have rounded windows.

**Adam Long**
Industrial design student
University of New South Wales
Canada Bay, Australia

# ❓ On the blink

*When I watch digital TV channels from terrestrial transmitters,
I have to endure periodic disruptions during which the audio and
images start stuttering. I recently realised that the disturbances
occur every time motorbikes – and particularly scooters – pass my
house. It doesn't happen with cars. How do scooters disrupt my
TV?*

**Michael Smith**
High Wycombe, Buckinghamshire, UK

Petrol engines use an electrically generated spark to ignite the
fuel-air mixture. A modern car uses a solid-state electronic
ignition system, connected to the spark plugs using cables
which reduce the level of electromagnetic radiation from the
engine.

In contrast, many motorcycles and scooters, especially
those with two-stroke engines, use magneto ignition systems.
A magneto is a simple mechanical device involving a coil
and a magnet that can generate a high voltage when it is
needed. It is connected to the spark plugs using metal cables
rather than the high-resistance cables used in cars, which are
not well suited to a magneto system. As a result, magneto
systems emit much higher levels of electromagnetic radiation
than electronic ignition systems.

The high-frequency component of this radiation can
interfere with broadcast signals to such an extent that your
receiver's error-correcting routines cannot cope.

**Stewart Haywood**
The Colony, Texas, US

This problem is worse when receiving terrestrial digital TV
than with analogue transmissions. Analogue technology is
far more robust and just displays such interference as white

dots on the screen. However, with digital technology there is a very high level of coding in the signal waveform. The result is that ignition interference (and indeed other kinds of interference) can disrupt the decoding process in the receiver, causing the picture to stutter or even freeze completely while the decoder recovers and gets back in sync.

Sometimes replacing the aerial cable with a double or triple-screened variety can cut the amount of ignition pick-up and so reduce glitches.

**Richard Harris**
Malvern, Worcestershire, UK

In the UK analogue TV transmissions are due to be switched off in 2012. Until then, digital terrestrial signals are being transmitted at reduced power to avoid interference with analogue signals. When the digital signal strength is eventually increased, the improved signal-to-noise ratio detected at the receiver should overcome this problem.

I should add that I cannot receive digital terrestrial signals at all. My neighbour can, and suffers from the same problems as your correspondent. We have identified the individual scooters, motorcycles and cars responsible – they are all rust buckets.

If it makes anyone feel better, I watch free-to-air satellite channels with a 1.2-metre dish and find that one particular motorcycle interferes with that too. This is amazing as the signals are all in the gigahertz rather than the megahertz range, and the interference is coming from outside of the line of sight to the satellite.

**Adrian Frame**
Norwich, Norfolk, UK

# Contrary cotton

*Some garments made of 100 per cent cotton will hang-dry on the washing line without creases, while other pure cotton items end up covered with stubborn creases that only a steam iron will shift. Why is this? Is it the quality of the cotton, or perhaps the species of cotton plant it comes from? And while we're on the subject, some cotton towels never really absorb the moisture from your body, however many times they are washed, while others – initially resistant while new – age into the job perfectly well after visiting the washing machine. What's going on?*

**Diana Ball**
By email, no address supplied

One of the reasons for a difference in crease retention is that cotton varies in its fineness. Fine cotton tends to form fewer creases than coarse fibre cotton. Also, a tight weave will tend to crease more than a loose weave, and woven fabrics will crease more than knitted fabrics.

Washing is responsible for most creases found in the fabric. When cellulose, an essential component of all plants, including cotton, gets wet, the hydrogen bonds holding it in shape are broken. If the fabric is dried in a creased state, new bonds form that hold these creases in place, where they stay, sometimes even after vigorous ironing.

Many cotton fabrics today receive chemical crease-resist-ant treatments. The presence, type or quality of this treatment is the likely reason you are experiencing major differences in the crease retention of cotton garments.

Regarding towels, one made with pile yarns consisting of fine fibres will dry you much better than a towel made with yarns consisting of coarse fibres. This is because fine fibres form a greater number of finer capillaries that wick the water away from the body more efficiently.

Years ago, all towels were made with loops of fibre, or terry, on both sides of the towel. Today, many are sold with looped terry on one side and a cut pile of single fibres on the other. The fibres on the side of the towel that has the cut-pile tend to separate, which reduces their water-attracting capillary action and by consequence their ability to dry your skin. So while the cut-pile side of the towel has a rich and velvety appearance, the looped terry side is much better for drying purposes.

**J. Robert Wagner**
Plymouth Meeting, Pennsylvania, US

A very high percentage of what is labelled 100 per cent cotton has been chemically treated to enhance the properties of the garment or linen. Much of this treatment is focused on modifying the surface of the fibre.

In the 1950s cotton fabric was treated with epichlorohydrin, which very effectively prevented wrinkling by keeping the individual fibres straight and a little springy. Unfortunately the compound weakened the fabric, so it was replaced by new treatments, including resin coatings. Many of these coatings are still in use today.

If you have ever visited a fabric shop, you may recall a pungent odour. This comes from the formaldehyde that is used to treat cotton fabrics and is present in a type of phenolic formaldehyde resin. The latest fabric surface treatments include Teflon resins to provide wrinkle resistance and prevent staining.

Some towels and bed sheets are treated with heavier coatings of polymers to give a very soft feel which, while pleasing, has the unintended consequence of repelling water. I personally discard any towels with this permanent softener treatment because they remind me of trying to dry myself with a plastic bag.

Incidentally, it would be prudent to thoroughly wash any new garment before wearing it and allowing it into contact with the skin. Many of the cotton treatments described above can be hazardous for those working with them during manufacture – for example, some fabric softeners can trigger allergic reactions.

**Charles Caban**
Smyrna, Georgia, US

## ? In the clink

*My father-in-law used to tape family mealtime conversations. When played back, the background noise – like silverware hitting plates and doors closing – is surprisingly prominent. Why is it that we filter these sounds out as they happen, but seem unable to filter them out when we listen to the recording?*

**Gary Yane**
Oldenburg, Indiana, US

Microphones are wonderfully objective devices. They detect variations in absolute pressure or the pressure gradient on a particular axis and faithfully transduce these into electrical signals. In contrast our ears have a brain attached, and between them they do a much more subjective job – interpreting our acoustic environment, not just recording it.

Our ears themselves are simple pressure transducers, but we also have ways of working out where sounds are coming from. For this we make use of the relative levels, phase and arrival times of sounds. In addition, the shape of our head distorts the local audio field in a way that we are personally familiar with, and this aids location of sound sources, particularly when we can move our heads.

We detect not only direct sounds but also reverberant ones. The space we are in significantly colours and adds to sounds, mostly in the form of delayed noise from random reflections. This would severely reduce the intelligibility of sound if the brain were not adept at adjusting to these conditions. It works out when the noises arrive, and where from, and can largely ignore them if it so chooses.

When the sound objectively recorded by a microphone is replayed through a single loudspeaker or even a stereophonic system, the random cutlery-clanging reverberant sound that should be all around us is now directly in front of us. Directional and timing cues that the brain would normally use for filtering are now inconsistent or just plain wrong.

**Chris Woolf**
Liskeard, Cornwall, UK

# ? Monkey business

*My son has a game in which you hang small plastic monkeys and gorillas from a plastic network of tree branches. The branch network is attached to an overhanging trunk by a magnet. As more monkeys are hooked on, the network becomes more unsteady until the magnet can hold it up no more. The player who breaks the bond is declared the loser. The game depends on a magnet of a clearly defined strength. But how is this strength determined so accurately during manufacture that it can hold almost all the monkeys (but not quite)?*

**Pavel Sypchenko**
Kiev, Ukraine

The strength of a magnet depends on its material, shape and preparation. When magnets are made these properties are fairly well standardised to within a few per cent of the strength required. For the purposes of this game, that is precise enough. You only need the game to last beyond the first few monkeys and end before you run out of monkeys.

Secondly, there is far greater variation in how the game is played than in how magnets are made. Magnets are exquisitely sensitive to how well they fit together. When polished to fit very closely they can cling to each other amazingly strongly, whereas the merest roughness, dent or sprinkling of dust can weaken their attraction by a large factor. Details of their orientation also affect their strength.

Players of the game are unlikely to pay much attention to such finicky details, so the strength of any set-up is likely to vary more in other ways than in the innate strength of the magnet. Then again, a really thoughtful player, with steady hands and a good eye for which plastic monkey to select and where to hang it, might survive several rounds longer than a slapdash player.

**Karl Kracken**
Hull, East Yorkshire, UK


Manufacturers are more likely to make a magnet and then test the number of monkeys it can hold before the network of branches will fall, than manufacture the magnet to a specific strength. This way they can supply a number of monkeys in the game box in excess of the magnet's strength. Supplying a few extra monkeys is a lot cheaper than going through weeks of R&D to create the perfect magnet.

**Sohil Patel**
London, UK

# ? Rank and dank

*Why do wet things smell more than dry ones?*

**Ben Scullion (age 4½)**
Darlington, Durham, UK

Making something wet does not automatically make it more smelly. For instance, a wet clean towel smells no worse than a dry clean towel. However, the presence of moisture does allow the growth of bacteria, assuming that there is organic matter present for the bacteria to eat. As they grow and multiply, bacteria produce a whole range of smelly compounds of the kind you can detect in bad breath, for example. So, given moisture and enough time for bacterial growth, wet things can smell worse. But if you prevented bacterial growth by, say, sterilising the wet item to kill all bacteria, then it wouldn't develop such a smell.

**Simon Iveson**
Department of Chemical Engineering
University of Newcastle
New South Wales, Australia

Most chemical components of dank smells are products of microbial activity, and microbial activity requires water. Once the chemicals are present they can reach the nose only by escaping into the air.

Most are fatty acids, amino compounds and the like, with charged chemical groups that readily bind to non-volatile molecules such as large proteins and carbohydrates. Once they have latched onto, say, dry cloth or leather, they cannot float freely into the air so there is not much to smell. However, these charged groups have an affinity for polar molecules, and the most polar of common molecules is water. So when the object gets wet, water molecules prise loose the odour molecules, cocooning them in tiny mobile parcels of water. For good or ill many escape into the air, reaching nearby noses in vast numbers.

Accordingly, a powerful deodorising strategy is to release other molecules that immobilise pong molecules by binding them with complementary charged groups. Chlorophyll combats smells partly by presenting a metal atom that binds the active groups of many smell molecules. Similarly, by binding key molecules, partly oxidised paraffin wax vapour from the smoke of burning candles also helps clear a room of the stench of cigarettes.

**Jon Richfield**
Somerset West, South Africa

## ? Blow hole

*What is the purpose of the small hole halfway down the outside shell of a Bic ballpoint pen?*

**Frank Horseman**
Derby, UK

If the inside of a ballpoint pen were entirely airtight, the pressure inside it would fall as the ink was used up. This would slow or stop the flow of ink because the higher air pressure outside the pen would push the ink back in.

The reverse could happen if the pen were heated. This would cause the ink to leak out of the pen (presumably onto your most expensive jacket). The hole is there to allow the air pressures to equalise and prevent these problems.

**Simon Langley**
Ilkley, West Yorkshire, UK

*Sven Taylor of Riedisheim, France, took the question to the manu-facturer's French factory. Here is the answer he received – Ed.*

The hole is to equalise the pressure inside the pen with the pressure outside the pen. These vents, or holes, in the pen barrels, basically help to prevent ink leakage. Approximately 90 per cent of all pens are vented to prevent leakage. Pens that do not have vented caps contain sealed ink systems and must be pressurised.

**Société Bic**
Communication Department
Clichy, France

# ❓ Shampoo pooh-pooh

*I have just read the label on my shampoo bottle. The list of ingredients is mind-boggling. How on earth did anyone come up with such a complex concoction and what exactly are chemicals such as sodium diethylene-triamine pentamethylene phosphonate and hydroxyisohexyl 3-cyclohexene carboxaldehyde doing to my hair?*

**Haydn Ford**
Hexham, Northumberland, UK

Sodium diethylene-triamine pentamethylene phosphonate is one of several chelating agents present in the shampoo. These are added because they can complex and deactivate metal ions, stabilising the shampoo against any degradation caused by these ions. They also complex the calcium and magnesium ions present in hard water, preventing the shampoo from separating out on the hair. In some special shampoos these agents can be used to remove the copper ions deposited on hair when swimming in chlorinated water, which can give a greenish tinge to fair hair.

Hydroxyisohexyl 3-cyclohexene carboxaldehyde is one of several fragrant materials which will be present in the shampoo. Together these produce a shampoo that smells attractive and leaves the hair with a fragrance created to make the user feel their hair has been cleansed and refreshed.

How does anyone come up with such a complex concoction? Shampoo formulations are not 'concocted'; each component is there for a purpose. Those developing such formulations are driven by the need to create a product that users will want to keep buying because they believe it is doing good to their hair.

**Tom Jackson**
Wigton, Cumbria, UK

Believe it or not, most of the ingredients in shampoo are doing nothing to your hair. Only the detergents clean it, while the rest of the ingredients are used to improve the appearance, smell, texture and shelf life of the product. Providing there are no restrictions on the use of an ingredient, manufacturers are free to use just about anything they like in a cosmetic or toiletry.

A typical shampoo is mostly water, containing between 5 and 20 per cent detergent, with shampoo for dry hair containing less detergent than shampoo for greasy hair. The most widely used detergent is sodium lauryl sulphate (SLS), but as this gives a poor lather, sodium laureth sulphate, which produces a stronger foam, or foam boosters such as cocamide DEA or cocamidopropyl betaine, may also be present. Lather plays no part in the detergency process, although it does keep the detergent and any other active ingredients close to your hair and scalp. Thick lather also reinforces the psychological link between the shampoo and perceived cleaning power.

Oils can be added to counteract the drying effect that detergents have on hair and, therefore, emulsifiers and emulsion stabilisers must also be added. The oils can be anything from natural vegetable oils to synthetic silicone polymers such as methicone and dimethicone. These also have a conditioning effect, helping to smooth the cuticle layer of the hair shafts.

In addition to oils, 'all-in-one' conditioning shampoos contain 'film formers' to distribute the oils, and antistatic agents, such as cationic surfactants or silicones, to reduce the build-up of static charge in your hair.

At least two preservatives are normally present. One must be water-soluble to protect the watery part of the shampoo, and the other oil-soluble to preserve the oils in the emulsion. The paraben family, formaldehyde, glyoxal and the methylchloroisothiazolinone/methylisothiazolinone mixture are all preservatives commonly added to shampoos.

Thickeners are added to adjust the texture and pouring properties of the product, colourants impart the required colour, UV-absorbers stop the colours from fading, opacifiers give the shampoo a creamy or pearlescent appearance, and natural or artificial fragrances make it smell nice. Shampoo for greasy hair might have a citrus fragrance because of the psychological link between lemon juice and grease-cleansing ability.

Specialised shampoos for dandruff treatment commonly contain zinc pyrithione, and shampoos for treating head lice or fleas contain insecticides. Shampoos to protect dyed hair from the bleaching effect of chlorine and chlorine-liberating compounds in swimming pools contain mild reducing agents, such as sodium thiosulphate, sodium sulphite or sodium nitrite. Once hair colour has faded, however, these shampoos cannot restore it.

Finally, the questioner's hydroxyisohexyl 3-cyclohexene carboxaldehyde, which also goes by the trade name Lyral, is an artificial fragrance, often used to mask unpleasant odours. Because it is a known contact allergen and can cause sensitisation, its use in cosmetics is regulated within the European Union.

**Steve Antczak**
Co-author of *Cosmetics Unmasked: Your family guide to safe cosmetics and allergy-free toiletries*
Lymington, Hampshire, UK

# ❓ Round the twist

*I have often observed defects, or knots, in helical (or spiral) telephone cords. It takes a considerable effort to untangle the cord, which seems to be the only way to restore uniform helicity. How do these defects happen and how can they occur spontaneously during normal handling of the telephone?*

**Jan Stumpel**
Tokyo, Japan

The telephone cord is 'set' into a helical form during manufacture. Despite its structure being twisted, the cord is stable in that configuration. When the user rotates the handset in the opposite way to the direction of the helix, some of the in-built twist is removed. However, the de-twisted cord does not become straight, but forms a length of helix in the opposite sense. The greater the reverse rotation of the handset, the greater the length of reverse helix. At the point where the two helixes meet, there is a short region where the twist direction reverses. This is the apparently hooked section which is irritating to telephone users. If one rotates the handset in the direction of the original helix, until all the reverse-twist has been removed, the reversal will soon disappear.

This phenomenon can be seen in yarns whose characteristics are modified in the false-twist textile process, which produces continuous-filament yarn. Twist is inserted into a bundle of filaments, the helical form being stabilised by heating. The twist is then removed. The filaments do not become straight; instead they form helixes of rapidly reversing sense, much like the telephone cord in question. The helixes give the yarn its characteristic stretch properties. Unlike the single telephone cord, textile yarns are usually multi-filament. The detail is therefore rather more complex, but the principle is the same. You can see the same effect under a microscope using an old pair of stretch tights.

**Graham Waters**
Pontypool, Gwent, UK

Knots in telephone cords are born as metastable knot-antiknot (k-a) pairs when the cord is locally twisted against its natural coil. With a little practice you can resolve a k-a pair into its components, and propagate one element (say the antiknot) to the end of a new cord, where it can be annihilated by twisting the handset. This adds a global twist to the lead and leaves a stable knot in the middle of the cord. In normal use, dormant global twists (gts) are induced by random movements of the handset. They are barely detectable against the natural background, since their only effect is to change the torsional energy of the cable and the total number of turns from end to end. The chance encounter of a dormant gt with a k-a pair (formed by an idle hand) then generates an apparently spontaneous and very stable knot. On a larger scale, you might create an entire observable universe by annihilating spontaneous antiparticles. God doesn't play dice, He fiddles with phone cords.

**Alan Calverd**
Bishop's Stortford, Hertfordshire, UK

The effect occurs spontaneously in my household because the receiver undergoes a half twist every time the phone is answered and a further half twist when the receiver is replaced – thus gradually uncoiling the wire until it can stand no more. Perhaps this is just another example of parity violation – some phones change helicity, others (like mine) don't. This could be a ground-breaking area in physics research – if only we could find a macroscopic equivalent of the top quark or tau neutrino!

**Mark Burbidge**
Birmingham, UK

## ? On the line

*How long a line could you draw with a single pencil?*
**Sulin Milne**
Newcastle Emlyn, Dyfed, UK

*Forget string – we now have a new saying, 'How long is a pencil line?' – Ed.*

My mind boggled at the range of variables implied by the question. As a scientist turned engineer (now retired), I decided to conduct a simple experiment in which the variables were reduced to a manageable number.

Selecting a clutch pencil with a 0.9-millimetre diameter, and using the manufacturer's H lead, I drew 100 lines each 30 centimetres long on high-quality printer paper. The pencil was tilted at 75 degrees to the plane of the paper. By measuring the reduction in the length of the refill and allowing for clutch-clamping wastage, I concluded that 541 metres of line could be drawn with a lead 60 millimetres long.

I also found by inspection that small changes in some variables had a large effect on the rate at which the lead was used up.

**Peter Peters**
Sherborne, Dorset, UK

Taking the simple case of a clutch pencil, I found by experiment that a 1-millimetre length of 0.5-millimetre 2B lead would draw about 9 metres of uniform line on ordinary photocopier paper. In my clutch pencil a new lead has a usable length of 50 millimetres, so that's 450 metres of line per lead. Looking at it another way, it's easy to work out that 1 cubic millimetre of pencil lead is needed to draw 45.84 metres of line.

A brand-new wooden pencil from a reputable maker is 175 millimetres long with a lead diameter of 2 millimetres. Assuming it is possible to use all but the last 20 millimetres of the lead, and (crudely) that each millimetre of lead draws 9 metres of line as with the clutch pencil, that would give us 1,395 metres of line for the whole pencil.

However, the volume of usable lead in the pencil, assuming again that the last 20 millimetres can't be used and that half is lost to sharpening, is 243.5 cubic millimetres. At the same volumetric wear rate as in the clutch pencil, that should produce 11,162 metres of line. I expect the actual output will be somewhere between these two answers.

The hardness of the lead will make a difference, as will paper type, the density of the line and how careful the user is not to sharpen too often or too far.

**Andrew Fogg**
Sandy, Bedfordshire, UK

# ❓ Field of bubbles

*When I placed a glass of still mineral water in front of my computer screen I noticed that tiny bubbles started to form around the edge of the glass. Why does this happen and is the water still OK for me to drink? And if it was the screen that caused this, what is happening to my body, which is essentially liquid, when I sit in front of my computer?*

**C. Strange**
Lancashire, UK

The correspondent who claims that bubbles form in mineral water as a result of him placing a glass of water in front of a computer screen is mistaken about the cause.

I always have a glass of water by my side. If I leave that glass for an hour or two, without touching it or drinking from it, small bubbles form over the glass surface below the water level.

I presume in your correspondent's case that the mineral water was bottled from a pressurised source where more air was dissolved than would be if it was bottled at atmospheric pressure. In my case, the bubbles come from the air dissolved due to the mains pressure in the pipes. The water is perfectly safe to drink, including its bubbles.

**Gordon Thompson**
Crewe, Cheshire, UK

Water has air dissolved in it, and the colder it is the more air is dissolved. Fish tanks found in aquariums and homes make use of this dissolved air to allow the fish to breathe. They must also have some means of refreshing the air content of the water, or the fish will suffocate and die.

Your glass of mineral water was probably chilled when you poured it out so a lot of air will have been dissolved in it. As it warmed up, air came out of solution and formed bubbles on the side of the glass. Being in front of the monitor may have warmed the water a bit more quickly, but those bubbles were going to appear anyway.

**Stephen Forbes**
Leeds, UK

# ? Humming words

*Electricity is supposed to be the 'silent servant'. So why do transformers hum?*

**D. J. Priestley**
University of Wales, Swansea, UK

Transformer hum is caused by a phenomenon known as magnetostriction. To understand why, it is necessary to take a look at how transformers work.

Inside they contain two coils of wire, the primary and the secondary coils, wound onto opposite sides of a ring made out of many thin sheets of iron or some other ferromagnetic material.

An alternating current flowing through the primary coil generates an alternating magnetic field in the iron ring, which in turn creates a voltage in the secondary coil. The ratio of the primary voltage to the secondary voltage is equal to the ratio of the number of turns of wire in the primary coil to the number of turns in the secondary. This allows us to change the hundreds of thousands of volts running through overhead power lines to a voltage low enough to be safe to use in our homes.

The iron making up the ring that joins the primary and secondary coils is divided into microscopic domains. In each of these domains, the magnetic field points haphazardly in different directions, much like a classroom full of unruly pupils who are running all over the place.

However, when the iron is placed in an external magnetic field, these domains tend to line up and add together, producing a strong magnetic field pointing in one direction, just as schoolchildren will snap to attention at a teacher's command.

As the domains line up, the material very slightly changes its length to accommodate the rearrangement. This is magnetostriction. As the magnetic field through the iron alternates, the iron expands and contracts over and over again. These vibrations produce the sound waves that create the transformer's distinctive hum.

In the US, the mains voltage alternates 60 times every second (60 hertz), so the material expands and contracts 120 times per second, producing notes at 120 Hz and its harmonics. In Europe, where the mains supply is 50 Hz, the hum is nearer 100 Hz and its harmonics.

**Michelangelo D'Agostino**
Berkeley, California, US

In addition to magnetostriction there are two other reasons why transformers tend to emit sound.

The first is imperfect insulation. Just as the corona discharge from power lines in damp air produces a buzzing sound, insulation breakdown in a transformer can also be noisy. In practice, however, insulation breakdown usually occurs deep inside a transformer, where the heat stress is most severe, and no audible noise emerges until the final catastrophic failure.

The second is caused by moving parts. Power supplies such as those you find behind computers sometimes make a buzzing sound, which is most likely to be the wire winding moving as the transformer's magnetic field and the current passing through it act together to produce a force similar to that in an electric motor. On the face of it, it seems that eventually metal fatigue ought to set in, but in practice transformers seem to be able to keep buzzing for years.

Other parts of a transformer can also buzz. For example, if the clamps that hold the parts together are not fixed tightly, they can rattle inside the casing.

**David Billin**
Carshalton, Surrey, UK

The description 'silent servant' did not really mean that electricity was silent. The expression was coined in the early 1920s by the General Electric Company in the US and used in advertisements and popular magazine articles to promote the use of electrical equipment in homes. The idea it was meant to convey was that electricity, unlike humans, could perform tasks without speaking or being spoken to, not that electricity itself was silent. Indeed, many pieces of electrical apparatus were, and still are, quite noisy.

**Michael Brady**
Asker, Norway

Silent transformers do exist and have been around since the early 1980s. The first models were too heavy and bulky for many types of equipment. But in modern appliances, 'switch-mode' power supplies are used which have much smaller transformers supplied by alternating current at a frequency too high for humans to hear, with sharp-edged pulses rather than smooth signals.

The mains AC frequency of 50 or 60 hertz – which is noisy when passed through a transformer – is increased to the higher frequency, usually via an oscillator. The current then enters one or more transformers that step down the voltage and, thanks to the inaudible nature of the higher frequency, allow the transformer to perform quietly.

The more rapidly changing magnetic field allows smaller transformers to be used. So as well as converting the audible humming sounds to inaudible, ultrasonic whistles, it helps make equipment smaller and lighter.

**Patrick McTiernan**
Swindon, Wiltshire, UK

## ? Let's twist again

*Given that the average person twists and turns up to 100 times during a night's sleep, why is it so unusual for anyone to fall out of bed? Does the human brain have a built-in warning system that is triggered when one's body goes near or over the edge?*

**G. Rowlands**
Purley, Surrey, UK

Some 25 years ago, at the University of Edinburgh, Geoffrey Walsh and I investigated the reasons why adults do not usually fall out of bed while asleep. Since no one can know what movements they carry out during sleep unless some form of recording is used, we devised a simple experiment.

Volunteers slept on a very wide mattress in a warm room, with no coverlet so that they would not be able to detect in their sleep where they were in the bed. Their head position was noted from a choice of four positions: nose to left, nose up, nose to right, or nose down. The apparatus was unsophisticated, and comprised a rugby scrum cap onto which I stitched a circle of plastic tubing complete with a short piece of glass tubing. The tubing contained some mercury, and I thrust some needles through the tubing wall at suitable points and attached a dry battery so that small voltages were generated according to the head position. These were recorded all night on an electroencephalographic recorder.

Periods of sleep and waking were recorded by arranging for a small sound to be made at around 10-minute intervals. If the volunteer was awake and heard it, they pressed a bell push attached to their clothing. This allowed us to discount movements made during this time. During sleep, of course, no response was recorded.

Participants turned at irregular intervals throughout their sleep, for example, nose to left, nose up, nose to right, then back again. But they never turned nose down. As a result, they did not roll over and over so that they would fall out of bed. Instead, they remained in roughly the same position all night.

At what age does this behaviour appear? Because we could not leave young children alone with such fascinating apparatus on their heads, I simply watched my niece and nephew, aged about four and two, in their cots over a period of some six hours while they slept. During the night they did turn nose down from time to time. So they could turn over and over, and could have fallen out if their cot sides had not stopped them.

I concluded that quite early in life we learn that it is difficult to breathe if we turn nose down, and we avoid it even when asleep. As a result we won't roll out of bed.

**John Forrester**
Edinburgh, UK


If you find not falling out of bed in your sleep an impressive skill, spare a thought for sailors. In some ships they still sleep in hammocks; and the naval version, called a mick, is slung tight and level. Though it does seem to reduce sensitivity to the ship's motion, any sleeper who cannot lie flat and still in a mick is a hostage to fortune. It dumps you instantly if you so much as breathe asymmetrically, and yet thousands of sailors have slept soundly in them for centuries.

**Antony David**
London, UK

*The ability of people to adapt to unfamiliar sleeping situations (from broad beds to narrow beds to hammocks to futons spread on a floor) suggests that on top of the processes investigated by John Forrester and Geoffrey Walsh we are somehow able to tell ourselves how much we can move before we fall asleep. In some ways this is similar to telling ourselves what time to wake up, which many people can do without an alarm clock – Ed.*

# ❓ Guilty as charged

*As back-up for my digital camera, I fully charged a set of AA nickel-metal hydride (NiMH) batteries, and carried them in a battery box with no chance of accidental connection. When I needed them some time later, they had completely discharged. Do rechargeable batteries leak their charge over time? If so, why, and how long does it take? For extra back-up, I now carry a set of ordinary alkaline batteries as well.*

**Joseph Oldaker**
Nuneaton, Warwickshire, UK

Nickel-metal hydride batteries have a high rate of self-discharge – about 30 per cent per month. This means that every two months their charge diminishes by a factor of 2, and in a year they will discharge to about 1.4 per cent of their full charge – effectively dead.

Nickel-cadmium batteries are somewhat better, and

lithium batteries are much better at only about 2 or 3 per cent discharge per month. In 2005, low self-discharge NiMH batteries were introduced. These are sold as pre-charged or ready-to-use. They are more expensive than old-fashioned NiMH batteries but can be used in many applications, such as a clock, where normal NiMH batteries would be unsuitable – though in some of these applications it might be more economical just to use non-rechargeable batteries.

**Eric Kvaalen**
La Courneuve, France

# 4 Plants and animals

## ❓ Jumping jumbos

*Is it true that elephants are the only quadrupeds that cannot jump?*

**Tad and Lydia Forty (aged 13 and 8)**
Bath, Avon, UK

*Thanks to Colin Watters and others for pointing out this splendid YouTube video of an elephant on a trampoline at bit.ly/1LWsJ – Ed.*

This is a fun question, but it is not true even if we include only four-legged animals that routinely walk on land.

Elephants cannot jump, from level ground anyway. This is true even when they are babies, as far as we know, but they are not alone. Probably all turtles cannot truly jump. It is also likely to be true for some salamanders and large crocodiles, some chameleons and other lizards. In fact, the statement is almost certainly not true even if restricted to mammals. Hippos probably cannot or do not jump, along with moles and other burrowing mammals, sloths, slow loris and other climbing specialists.

However, the truth is that no researchers have looked at this question in a rigorous way. We don't even know specifically why – in terms of detailed anatomical mechanisms and physics – any of these animals cannot jump. There are just scattered anecdotes and folklore, like the tired myth that elephants have four knees, which I still encounter again and again from the public. Elephants actually have two knees like all other mammals because their anatomy is essentially the same.

So the question is certainly worth addressing. But there are a lot of species out there, so as a general rule it's probably best to assume there is unlikely to be any species that is alone in being unable to do some seemingly common activity.

**John R. Hutchinson**
Reader in evolutionary biomechanics
Royal Veterinary College
University of London, UK

Racehorses weighing about half a tonne are among the largest quadrupeds that can make impressive jumps. In horse racing, the Chair, the highest fence on the Grand National course, is 1.8 metres high.

The largest wild animal I have seen making an impressive jump was an eland, one of a group that I saw galloping in Kenya. Its jump was high enough to have cleared the back of another eland, roughly 1.4 metres from the ground. The animal probably weighed about the same as a racehorse.

Large male African elephants weigh around 5 tonnes, and Asian elephants only a little less. After them, the heaviest quadrupeds are the hippopotamus (about 3 tonnes) and the white and Indian rhinos (about 2 tonnes). Whether these and other large animals can jump depends on what you count as jumping. A film I took of a white rhino galloping at 7.5 metres per second showed that, at one stage of its stride, all four feet were off the ground. I do not think of that as jumping, but I cannot think of any clear-cut definition of jumping that would exclude it.

Big jumps require strong leg bones and muscles. The vertical component of the force the feet exert on the ground, averaged over a complete stride or jump, must equal the animal's weight. In a substantial jump, the animal is off the ground for longer than it would be in a running stride, so its legs will be subject to larger forces at take-off and landing.

Simple physics tells us that if big animals were precisely scaled-up versions of smaller ones, their weights would be proportional to the cubes of their linear dimensions. The cross-sectional areas of bones and muscles, however, would be proportional only to the squares. An animal with double the linear dimensions of another would be eight times as heavy, but its legs would be only four times as strong, and so less able to jump.

Of course, even closely related animals of different sizes are not scale models of each other. For example, a 500-kilogram eland has relatively thicker, straighter legs than a 5-kilogram dik-dik – but the differences are not sufficient to eliminate the disadvantage for large jumpers.

Other than size, a quadruped's anatomy or physiology may be unsuitable for jumping. Some desert lizards that burrow in loose sand have greatly reduced limbs, tortoises have very slow muscles and the limbs of moles are highly modified for digging. I have never seen any of those quadrupeds jump, and do not expect to.

**R. McNeill Alexander**
Emeritus Professor of Zoology
University of Leeds
West Yorkshire, UK


Elephants are not the only quadrupeds that cannot jump. Some of the quadruped dinosaurs could not jump, including apatosaurus and diplodocus.

**Edward Rivers (aged 7)**
Angmering, West Sussex, UK


Really heavy animals like rhinos and hippos can hardly jump

or land without injury. After reaching terminal velocity, mice would bounce after hitting the ground whereas elephants would break, or, according to urban legend, 'splash'.

Even so, don't jump to optimistic conclusions if a large animal chases you over a ditch. Does it still count as 'being able to jump' if the jump causes the animal injury? If so, then you are in trouble because, yes, Indian elephants can jump. J. H. Williams in his book *Elephant Bill* relates how a stampeding female jumped a ditch handily, though she went lame in both forefeet soon after.

**Jon Richfield**
Somerset West, South Africa

## ❓ Water, water, everywhere

*One of your previous books has explained how fish drink. But what about water-dwelling mammals such as dolphins and whales. Do they get thirsty? And if they do, how do they drink?*

**Daniel Gough**
Glasgow, UK

Dolphins and whales do not drink. Just as we humans cannot use salt water as our source of water, neither can marine mammals. This is because they would need to ingest more fresh water than the seawater they consume in order to excrete the salt it contains.

Much of their water comes from fish and squid, which can contain more than 80 per cent water by mass. They can also obtain water through metabolising fat. In order to reduce their water loss they have similar internal designs to those of desert-dwelling mammals, including a long loop of Henle in the kidney nephron.

As well as internal adaptations, marine mammals did away with sweat glands to stop any water loss through sweating. Instead, they use their surroundings to cool down.

**Matthew Tranter**
Newcastle-under-Lyme, Staffordshire, UK

Marine mammals certainly are less prone to thirst than land-dwelling mammals; for one thing, they have no need to sweat. They do not swallow any more salt water than they can help, though. Unlike seabirds and turtles, they lack special salt-excreting glands, so every bit of salt they swallow exacts a penalty.

However, whales eat animals, and sirenians (manatees and dugongs) eat plants. In such foods, salt is as little as one-fifth as concentrated as in seawater because the food target has expended energy to excrete salt.

You might say that marine mammals rely on their food to desalinate their water. Even mammalian prey can contribute to this process as they eat low-salt organisms.

Interestingly, as they are unable to sweat or increase their water intake dramatically when thirsty, whales and seals are vulnerable to changes, especially increases in water temperature. In particular, many species have great difficulty crossing the equator, while most are comfortable as close to the poles as foraging will take them.

**Jon Richfield**
Somerset West, South Africa

# ❓ Foil attack

*I get two bottles of milk delivered to my house each day. One,
containing whole milk, has a silver foil top, whereas the other,
containing semi-skimmed milk, has a silver top overprinted with
red stripes. Based on observation over several years, the local
magpie population will often try to peck at and remove the striped
top but hardly ever attack the plain silver foil top. Have other
readers observed magpies or other birds being so discerning, and is
there a scientific explanation for it?*

**Barry Chambers**
University of Sheffield, UK

There are two potential explanations. Firstly, the magpies in
your garden may be showing some form of aversion to the
silver bottle-tops. Birds often show unlearned aversions to
food of certain colours, but these tend to be colours that are
associated with toxic insects, such as the black and yellow
stripes of wasps or the red and black spots of ladybirds. While
this could be the case with your magpies, I doubt it because
the red-striped tops would appear to be more reminiscent of
the colour patterns of toxic insects than the silver tops.

The theory I favour is that your birds know what's good
for them! Insectivores such as magpies need a diet rich in
protein with lower levels of both carbohydrates and fat. Just
like in humans, high-fat diets can cause magpies to suffer
from high levels of cholesterol and all the medical problems
associated with that. Birds are excellent judges of the toxin
and nutrient content of the food they eat and by choosing to
drink the semi-skimmed milk over the whole milk they get
the benefit of a high-protein food source without the costs
associated with eating fatty foods.

Maybe magpies could teach us all a thing or two about healthy eating.

**John Skelhorn**
The Institute of Neuroscience
Newcastle University, UK

In my youth, we used to have ordinary milk, with a red top, and creamy Guernsey milk, with a silver top, delivered to our home. The blue tits always attacked the creamier Guernsey bottles. After a couple of years, the dairy changed the cap colours to silver and gold, respectively. The birds learned the new colour code in about two weeks.

Your magpies are either stupid or fashionable – assuming, of course, there is a difference.

**Alan Calverd,**
Bishop's Stortford, Hertfordshire, UK

# ? Tomato attack

*Every time I collect tomatoes in the garden, my hands end up covered with an invisible substance with a pungent smell. It seems to come from the tomato leaves and branches. When I wash my hands with soap, the substance becomes very bright yellow-green – almost fluorescent – and it stains my soap, towel and wash basin. However, if I don't use soap to wash the substance off, it remains invisible. What is it?*

**Alex Saragosa**
Terranuova, Italy

Leaves of plants in the family Solanaceae (including tobacco, tomato, potato and capsicum) all have minute hairs on their surface which exude drops of a sticky fluid.

The function of this sticky substance is not altogether clear, but it could ward off attacks from aphids and other sap-sucking insects. The flavonoids and possibly other pigments in the fluid react with soap, which is alkaline, and change colour accordingly.

**David Whitehead**
Cape Town, South Africa

I remember being impressed long ago by the bright green colour that developed when washing my hands after helping my father tend his tomatoes. Some time later, as a botany student, I examined the tomato leaf epidermis under a stereo-microscope, with interesting results.

The tomato epidermis carries two types of multicellular hair – long ones of several millimetres readily seen with the naked eye, and much shorter hairs with four glandular sacs like short sausages at the apex. These are filled with khaki-coloured contents and have very thin, fragile cell walls. I found that prodding one of these sacs with a needle released sticky contents that could be drawn out as a thin thread which set rigid within about 2 seconds, leaving a deposit on the needle.

Later still, I found some photos of a mite wearing what can only be described as concrete boots. These were appar-ently made up of accumulated secretions from a tomato plant, collected as it rambled over the plant's surface. The deposit clearly encased the mite's legs, preventing it from hanging onto the leaf, thereby acting as a defence mechanism against this and other small, walking herbivores.

Casual brushing against a tomato leaf will transfer only a little of the substance. This reserves the secretion for organisms that ramble among the hairs – or anything grasping the plant strongly enough to bend the longer hairs.

The deposit's colour is lost against the skin of the average gardener, but it can certainly be detected by smell and, I suspect, by texture.

**Jim Kent**
Minehead, Somerset, UK


# ？ Quackers

*I was watching a duck and her eight chicks walking in a line across the grass. All of a sudden a couple of other chicks entered the group. The mother duck immediately weeded out the stranger chicks and sent them on their way. To us they looked identical, so just how did the mother duck achieve her feat? Is it just that animals are exquisitely sensitive to visual differences between members of their own species? Or was the mother duck relying on non-visual information as well and, if so, what?*

**Byung O Ho**
San Jose, California, US


Chasing away non-descendant young is called 'kin discrimination' and is often considered less efficient in birds than in other animals. However, eider ducks have been reported to discriminate against ducks that are not part of their family unit. Coots have also been seen to do the same thing, but neither species seems to use appearance as the way to recognise their young.

Many birds use acoustic recognition and can identify each other's voices. Swallows, finches, budgies, gulls, flamingos, terns, penguins and other birds that live in larger flocks do this. Odours can also play a role in determining how some birds recognise each other.

In ducks, sound seems to be the principal method of recognition: they have been fooled into returning to the wrong nest, only to be greeted by a portable cassette player rather than their ducklings.

The ability to recognise their own young saves colony-living birds from expending energy in raising someone else's offspring. It also stops ducklings running the risk of aggression from adults if they beg food from the wrong ones. Natural selection favours individuals who know who they are talking to.

Waterfowl have long been thought to be unable to keep track of their own young. They have been seen to lose their own ducklings to another parent, or to mistakenly accept and care for non-descendant ducklings. This has been put down to the fact that birds do not generally have a central family unit.

Ducks do behave in a different way towards their own ducklings, though. Parents sometimes favour their own offspring over non-descendant young, as with the duck in this question, or they may tolerate or encourage the ducklings to mix. Consequently, some provide what is called alloparental care, a form of adoption. This is seen when a duck is able to increase the chances of survival of her own offspring by accepting non-descendant ducklings into her entourage. Her own ducklings might be better off because the risk of any individual being eaten by a predator is lower if it is part of a bigger group. To improve the advantage even more, the non-descendant ducklings may be positioned at the edge of the brood, further away from parents. This has been seen in Canada geese; the adopted goslings were noted to generally potter further away from their adoptive parents than the biological offspring, and therefore not as many survived.

**Jo Burgess**
Department of Biological Sciences
Rhodes University, Grahamstown
South Africa

## ? Worm baiting

*While sitting on a bench beside a local green, I noticed a gull performing an excellent version of Riverdance. Then it stopped and scrutinised the grass around its feet. This sequence was repeated for about 15 minutes. I assume the gull was trying to attract worms to the surface with its rhythmic dance. Was it? If so, how does the strategy work?*

**Danny Hunter**
Dublin, Ireland

Yes, like many species of birds, some gulls have learned the earthworm-raising trick. Earthworms stay underground during the day unless flooded out by rainwater or alarmed by ground vibrations that suggest the approach of a mole. Just jab a garden fork into earth well populated with earthworms and some will pop out to avoid the little creature in black velvet.

Different birds have different techniques. Blacksmith plovers, rather than hunting earthworms, flush out grasshoppers, caterpillars and moths by tickling short grass with a trembling foot held forward.

Gulls that have learned the trick stamp for earthworms. Similarly, I have seen thrushes stamp by hopping hard with stiff legs. Once I was startled to see a red-winged starling watching an olive thrush's technique attentively, then having a go itself. It did get a worm or two, but its technique was faulty, with long, loping leaps instead of jerky thumps, so it did not scare enough worms and soon gave up. Or maybe it just didn't like the flavour of those earthworms it had caught.

**Antony David**
London, UK

The gull was indeed trying to get worms to surface. Underground, the rhythm of the gull's feet sounds much like rain.

Earthworms like to surface during rain because it enables them to move around overground without drying out – this is impossible when it is dry. By tricking the earthworms, the gulls get an easy meal. The gulls may have learned this trick from watching other gulls, or may have inherited the behaviour.

**Laura Still**
Devon, UK


I was sitting on Henley Beach in South Australia recently, watching a gull 'puddling' the sand at the water's edge before inspecting the water for any food items it might have disturbed. It seemed to be doing quite well for itself.

It appears that the gull seen by the questioner was applying successful food-gathering behaviour that evolved in one environment to another. This does make sense when one considers that gulls originate not in marine environments, as is frequently supposed, but in moorland ones. Presumably the behaviour evolved in environments that contained bogs, where the puddling behaviour would work successfully in damp ground some way from the water's edge.

The vital question, though, is whether the gull was successful in drawing up worms, or anything else edible, to the surface.

**Graham Houghton**
Aldgate, South Australia

## ❓ The hole truth

*I have always been fascinated by evolution, and while I can usually see why and how certain characteristics evolved in different species, I'm confused by whales and dolphins. How did their breathing holes evolve, bearing in mind their ancestors were land mammals?*

**Joe Bilsborough**
Tarbock, Merseyside, UK

Blowholes are paired nostrils that evolution has shortened and redirected towards the most convenient spot for snorkelling – the top of the head. They do not pass through the brain, though.

As in most swimming, air-breathing vertebrates – such as frogs, crocodiles, capybaras or hippos – whales' nasal openings, or nares, are placed high up so they can breathe with as little raising of the head or snout as possible. They also have protective valves to keep water out.

However, most of the creatures in that list are oriented largely towards the world above: they periodically leave the water for terrestrial activities and they float with nostrils and eyes just above the surface, watching for food and threats.

In contrast, the terrestrial ancestors of ichthyosaurs, cetaceans and sirenians (manatees and dugongs) evolved into creatures with their attention directed towards the underwater world. Their ears and eyes did not migrate upwards, only their nostrils shortened and the nares migrated towards the highest part of the head because although food and threats no longer came from above, the air they needed to breathe still did. In sirenians that migration is incomplete, so watch this space for another 10 million years.

**Antony David**
London, UK

The benefit of the location of cetaceans' blowholes is clear, but it's not so clear what factors motivated the initial steps in the migration of the nostrils from the nose to the top of the head. Natural selection certainly does not seem to have made significant progress until the whale's distant ancestors had irreversibly abandoned the land and shifted to a marine lifestyle.

The earliest identified precursor of modern cetaceans is Pakicetus, which lived during the early Eocene, about 53 million years ago. It resembled a hyena with hooves, was quite definitely a terrestrial animal and had nostrils at the extreme front of its long snout. It was not until the late Eocene – about 20 million years later – that the first 'true' whale appears in the fossil record. Named Basilosaurus, it featured nostrils that had migrated up its snout to a point just in front of its eyes. Basilosaurus had nostrils not only shifted backwards in comparison with its forebears, but also converging towards a location on top of the skull, in a clear move towards the modern arrangement. Because Basilosaurus was fully aquatic, it seems clear that it was the benefits of the modern set-up that were the driving force behind this particular aspect of its evolution.

Certainly, by the mid-Miocene, some 15 million years ago, the first modern whales and early dolphins all sported blowholes precisely where they are found on present-day species, although even today evolution has not arrived at a definitive form for a whale's nostrils. Baleen whales, such as the humpback and the blue, have two, while toothed species such as the sperm whale have just one.

Mystery still surrounds the reptilian predecessors in the cetacean's ecological niche. The dolphin-like ichthyosaurs were, as their classical name 'fish-lizard' suggests, particularly well developed for a marine lifestyle. They survived for 140 million years, almost three times as long as whales and dolphins have had to evolve from their terrestrial ancestors.

Yet even the very largest of ichthyosaurs retained two conventional nostrils set just in front of their eyes, very similar to Basilosaurus, despite this configuration requiring them to lift most of their heads out of the water to breathe, and exposing them to attack by predators.

So perhaps the real question is not why whales and dolphins have evolved their manifestly beneficial breathing arrangement, but why their reptilian analogues – and other marine mammals such as the dugong – did not do likewise.

**Hadrian Jeffs**
Norwich, Norfolk, UK

## ? The birds

*In New Zealand one of our radio stations broadcasts native birdsong each morning. It is obvious that seabirds have a much harsher screeching sound than the more melodious bush and land-based birds. In fact, I can usually tell a bird's habitat simply by the sound it makes. Why is there such a difference, and is it the same throughout the world?*

**John Finlayson**
Maungaturoto, New Zealand

Birdsong indeed varies by habitat type because the habitat has a profound effect on how these long-distance signals are transmitted. To minimise habitat-induced degradation, the acoustic adaptation hypothesis predicts that birds living in dense forests will have slower and more tonal calls, while those living in more open habitats will have faster-paced and buzzier calls.

The effect is most pronounced when comparing contrasting habitat types, such as very open and very closed ones.

Other factors, including the songs of species competing for acoustic space and the songs produced by closely related species, can also play a role.

**Daniel T. Blumstein**
Department of Ecology and Evolutionary Biology
University of California
Los Angeles, US

The subject is more complex than the question suggests. The South African bush hosts croaking corvids, harmonising antiphonal shrikes, shrieking parrots, raucous francolin, swizzling weavers and tweeting wagtails.

Calls seem to be adapted to distance, noise, obstacles, habit and competition. The most elaborate singers inhabit open bush, where their song can convey complex information over long distances. In thick bush, only deep ventriloqual notes such as those of the ground hornbill carry for any distance. White-eyes foraging among dense leaves cheep softly, keeping flocks together at short range.

Even the apparently unsophisticated croaks, screams and yarps of seabirds vary in complexity and carrying power according to their habits and individual circumstances. When calling through wave noise over long distances they tend to screech shrilly, whereas when they are intimate they are quieter.

Details vary, but the fundamental principles of auditory information encoding and transfer seem inescapable.

**Jon Richfield**
Somerset West, South Africa

# ❓ Balanced lifestyle

*Why do some birds stand on one leg?*

**Alexander Middleton**
Moorooka, Queensland, Australia

*Thanks to all those who offered the answer: 'If they picked up the other leg they'd fall over.' The old jokes are still the best – Ed.*

It has been proposed that the reason that flamingos stand on one leg is so ducks don't swim into them as often! The most likely answer, though, has to do with energy conservation. In cold weather, birds can lose a lot of heat through their legs because the blood vessels there are close to the surface. To reduce this, many species have a counter-current system of intertwined blood vessels so that blood from the body warms the cooler blood returning from the feet. Keeping one leg tucked inside their feathers and close to the warm body is another strategy to reduce heat loss.

I imagine the converse is true in hot climates – blood in the legs will heat up quickly, so keeping one leg close to the body will reduce this effect and help the birds to maintain a stable body temperature.

Another factor in long-legged birds is that it may require significant work to pump blood back up the leg through narrow capillaries. Keeping the leg at a level closer to the heart may reduce this workload.

It is also worth remembering that birds' legs are articulated differently to ours; what looks like the knee is in fact more like our ankle. Many birds have a mechanism to 'lock' the leg straight, so for them it is much easier to stand for hours on end on just one leg – on numerous occasions I have seen birds take off, and even land, on one leg.

**Rob Robinson**, Senior population biologist
British Trust for Ornithology, Thetford, Norfolk, UK

# ❓ Do the twist

*All the stems of the morning glory plants growing on my balcony coil in the same direction. When I moved some of the plants, I re-coiled them by hand onto the strings they creep around. Those that I had coiled in the 'wrong' direction started to coil in the 'right' direction as soon as they could. Why is this?*

**Judit Zádor**
Budapest, Hungary

Some winding plants such as morning glory and wisteria wind counter-clockwise (CCW). Others, such as hops and honeysuckle, wind clockwise (CW). Supposedly, you should not force them to wind in the 'wrong' direction or they will wither.

Although it is said that hops wind CW to follow the sun, the actual direction of winding is determined by the plant's genes and the pull of gravity.

A Japanese team at Kobe University led by T. Hashimoto chemically mutated straight-growing vines until some wound CCW, then looked at the molecular structure of the twisty bits. They found that a slight change in the structure of tubulin, a microtubule protein in cells, determined the winding direction (*Nature*, vol. 417, p. 193).

Another Japanese team, this time led by Daisuke Kitazawa at Tohoku University, found that gravity-sensing cells are crucial for shoot circumnutation – the bending and bowing of the tip – and the winding response (*Proceedings of the National Academy of Sciences*, vol. 102, p. 18742).

So a plant's gravity sensors tell it which way is up, and its tubulin structure determines whether it winds CW or CCW in relation to the vertical.

**Quinn Smithwick**
Cambridge, Massachusetts, US

# ❓ It's a dog's life

*Why do dogs like jumping into cold ponds, while cats and humans generally do not?*

**James Scowen**
London, UK

Your questioner appears to be confusing willingness with enjoyment. Most dogs are prepared to dive into cold water, but they may not like the experience. And in referring to cats, your questioner is almost certainly referring to the domesticated species, which is not necessarily representative of its genus.

Nonetheless, the canine tolerance for cold water, and feline intolerance, lie in their respective evolutionary histories. The dog (*Canis lupus familiaris*) originated in central Asia during the aftermath of the last ice age, at least 15,000 years ago. It is descended from the grey wolf (*Canis lupus*), with all the evolutionary baggage that implies. Ice-age wolves preyed on sub-Arctic herd animals such as elk, reindeer and caribou, which would have migrated in search of better grazing, crossing fast-flowing rivers swollen by meltwater when required.

Any animal – including the ice-age wolves – fording or swimming these rivers would have had to develop considerable physical and psychological resistance to low temperatures. Those that weren't prepared to get their feet wet wouldn't have lasted long enough to pass on their genes. Those that did bequeathed their doggy descendants a tolerance for cold water.

Some 5,000 years after the big bad wolf began the transition to being man's best friend, a group of wild cats (*Felis sylvestris*) in what is now western Asia apparently attached themselves to the local human population in a semi-symbiotic relationship. Significantly, the closest living relative of the proto-kitties is believed to be the sand cat (*Felis margarita*), a denizen of regions of extreme heat and aridity, such as the Sahara.

This ancestry was never likely to cultivate a hereditary tolerance for getting wet, even if natural selection had not already instilled a wariness of bodies of water, whatever their temperature. Large mammals have no freshwater predators in the sub-Arctic, but animals originating in the tropics have good reasons for not going into the water, most of them possessing very powerful jaws. A prehistoric African water hole was a fast-food outlet for large predators, both in and around the water.

The behavioural heritage of these widely differing ancestries can be most clearly observed when our modern-day pets are drinking. A dog will generally lap up its water enthusiastically, albeit with the occasional sideways glance at any animal that could attack. A cat, on the other hand, displays far more caution, constantly looking around suspiciously and keeping its body as far back as possible from the liquid.

**Hadrian Jeffs**
Norwich, Norfolk, UK

Dogs, like humans and cats, exhibit a homeothermic mode of temperature regulation – their body temperature remains constant in spite of fluctuations in the temperature of their environment.

Dogs are covered with thick hair to conserve internal heat, and regulate their body temperature through panting, an extremely efficient method. On a hot day it is quite common to see a dog with its mouth wide open and tongue hanging out.

Recent research has also indicated the presence of a complex network of blood vessels in the basal part of a dog's neck. This region functions as an efficient temperature regulator. In addition, dogs have relatively large spleens. When a dog is active or under stress, the spleen contracts and releases blood into the circulatory system, which provides yet another mechanism for carrying excess heat to the skin.

All this means that dogs are better adapted than humans or cats to withstand cold shocks or hypothermia.

**Saikat Basu**
Lethbridge, Canada

## ? Flight of the butterfly

*My 4-year-old daughter asked me how high butterflies fly. I was stumped. Can anyone tell us?*

**Jacque (and Tara) Lawlor**
Chelmsford, Essex, UK

Unlike humans, butterflies are not disposed to seeking altitude records. Indeed, they will not fly higher than is strictly necessary in their everyday lives, whether looking for a mate, food or somewhere to lay eggs, avoiding predators or migrating.

Worldwide there are many thousands of species of butterfly, each adapted to its own particular habitat and needs. Some spend their whole lives on a patch of coastal grassland, the larvae feeding on low plants or living in ants' nests, and the adults never flying more than a few feet above the ground. Others spend all their time in the tree canopy many metres above ground level.

Still others are only found on high mountains. So even though they don't actually fly very high above the ground locally, butterflies that live on the mountains of Peru spend their whole lives at altitudes of around 6,000 metres.

Butterflies that migrate tend to fly the highest in general. The most famous migratory butterfly is probably the monarch, *Danaus plexippus*. These leave Mexico each year and fly north to Canada, albeit taking several generations to get there. Monarchs have been sighted by glider pilots flying as high as 1,200 metres. Interestingly, they seem to fly in the same way as a glider, using updrafts to gain sufficient altitude so that they can glide for quite a distance before needing to use energy to climb again.

Europe also has plenty of migratory species. The painted lady, *Vanessa cardui*, makes its way to southern France from north Africa. It has to leave Europe in winter as no development stage of this insect can survive a frost.

To get to France many will cross through the mountain passes of the Pyrenees, which in general lie at about 2,500 metres. During late summer and autumn one can observe butterflies drifting southwards. If they encounter a high building, they just fly straight upwards and over it. If they encounter a high mountain range, they will do the same. So you need only to stand for a while on any mountain pass during the migration period to see them coming over either singly or in swarms, flying close to the ground as they travel.

The mountain passes of the Caucasus are higher, while those of the Himalayas are higher still at 7,500 metres. I wouldn't be surprised if migratory butterflies could fly straight over Everest if they encountered it in good weather.

However, insects of any kind cannot fly if they are too cold. Butterflies can keep warm to a certain extent by beating their wings, though if they fly too high in the wrong conditions, they may become too chilled to maintain a wingbeat.

On average, the air temperature reaches freezing at an altitude of just below 8,000 metres, suggesting that this would be their physical altitude limit. They might on occasion be carried higher on updrafts, but this surely doesn't count as autonomous flight.

**Terence Hollingworth**
Blagnac, France

The greatest acknowledged height achieved by migrating butterflies is 5,791 metres, set by a flock of small tortoise-shells, *Aglais urticae*, crossing the Zemu glacier in the eastern Himalayan mountains.

Not only is this an altitude record for butterflies, it is also the highest that any insect has been observed in controlled flight, comfortably exceeding the more frequent altitudes of between 3,000 and 4,000 metres at which monarch butterflies have been sighted by commercial airline pilots.

**Hadrian Jeffs**
Norwich, Norfolk, UK

## ? Brainy breeding

*Dog breeding often gets a bad press, including the apparently unfounded assertion that breeding for looks has an adverse effect on intelligence in dogs. But has anyone ever bred dogs, or any other species, purely for intelligence? Just how intelligent could any species get through selective breeding? And how quickly?*

**John Schofield**
London, UK

Intensive breeding for looks in any animal adversely affects intelligence and every other attribute – eventually including those very looks. This is intrinsic to selection, whether natural or artificial.

The effectiveness of selection depends on the range of

relevant genes in the population: the larger the natural population, the greater the range of genes is likely to be. Selection for any desired attribute rapidly reduces that range: in a single generation, less than 1 per cent of a population might be selected, immediately reducing the range of 'irrelevant' genes, including genes for mental or physical health or functionality.

Dogs bred for show are commonly selected so obsessively that any harmful genes they carry become fixed in their populations. In competitive show breeding, selection is particularly stringent, with the result that gene pools shrink rapidly. Most mutations and recessive genes in small, closed populations are harmful, so progress is overwhelmingly negative.

The closest we come to breeding for intelligence and functionality in dogs is in certain working breeds. Breeding companion animals specifically for desirable behaviour, intelligence and health should be gratifying, but it is also challenging and commercially precarious. People who need companions prefer to buy mongrels.

**Jon Richfield**
Somerset West, South Africa

Asking if anyone has bred a variety of dog purely for intelligence begs the question of what is meant by intelligence. The psychologist Robert Sternberg has shown that what we think of as 'intelligent' depends on what we value – specifically what we think people should be good at. So what we consider to be a clever dog would be one that does a good job at what we want it to do: herd sheep well or guard the house effectively. We have no need for dogs that are adept at calculus or playing the futures market, so we have never tested our capacity to breed this into them.

Sternberg identified three signs of intelligence: the ability to adapt to environments, the ability to shape environments and the ability to understand that the environment is not optimal, thus facilitating a move to a more congenial niche. On this basis, you could make the argument that almost all species are intelligent, even bacteria, because at the very least they are adapted to their environments. In addition, many can up sticks when things are not so good and move elsewhere, and some can even shape their environment in some way congenial to them.

Maybe dogs deserve special mention because they have shaped their environment by making themselves useful and appealing to humans, in return for food and shelter.

**Catherine Scott**
Surrey Hills, Victoria, Australia

## ? In the green

*The benefits of camouflage would suggest that there should be green mammals. Are there any – and if not, why not?*

**A. C. Henderson**
Braco, Tayside, UK

There is only one green mammal, the three-toed sloth. This is because a coat of algae covers the sloth's fur. Because of the sloth's tardiness and lack of personal hygiene, this is never cleaned off. No known mammal is capable of producing its own green epidermal pigment. The main reason for the absence of green mammals seems to be an ecological one. In general, mammals are simply too big to use a single colour for camouflage as there are no blocks of green large enough to conceal them. Most mammals have an environment that is made up of patches of light and dark and composed of many different colours. This means that those mammals which are camouflaged tend to be dappled or striped. Animals that do use green coloration for camouflage, such as frogs and lizards, are small enough to use solid blocks of green – leaves and foliage – for cover.

**Paul Barrett**
Department of Earth Sciences
University of Cambridge, UK

The main predators of most mammals are other mammals, especially the carnivores, such as the cat, dog and weasel families. Carnivores are all colour-blind or, at best, have very limited colour vision. Hence effective camouflage against them is not a matter of coloration but of a combination of factors such as brightness, texture, pattern and movement.

**Graeme Ruxton**
Scottish Agricultural Statistics Service
Edinburgh, UK

Your answers to this question only mention the tree sloth, which is not truly green, just covered with algae. There is actually a real green mammal – the green ringtail possum (*Pseudocheirus archeri*) – and what a lovely animal it is. The possum is a marsupial endemic to a small area in northeastern Australia. You can see a fine colour portrait of it in *The Complete Book of Australian Mammals* (Angus and Robertson, 1983). The article accompanying the portrait is by J. W. Winter, an expert on the possums of that part of Australia. He writes: 'This remarkably beautiful ringtail is aptly named: a mixture of black, grey, yellow and white hairs confers a most unusual lime-green colour to its thick, soft fur.'

I would add that it is also the most docile wild animal I have ever encountered. A scientist studying possums during the 1960s who caught one and kept it for a day before returning it to the wild allowed me to photograph it. It made no attempt to struggle, scratch or bite when taken out of the cage, nor did it try to escape.

Winter makes no comment as to whether the green colour has any apparent advantages, but he does report that 'its daytime roost, unlike that of other possums, is usually on an open branch. It sleeps upright, curled into a tight ball, gripping the branch with one or both hind feet and sitting on the base of its coiled tail, with the forefeet, face and tip of the tail tucked tightly into its belly.'

A motionless, amorphous green ball among the multitudinous shades of green in the rainforest would be far from obvious. The only predators Winter reports (apart from Aboriginal humans in the past) are nocturnal: the rufous owl (*Ninox rufa*) and the spotted-tailed quoll (*Dasyurus maculatus*). The latter is a marsupial carnivore, with a head and body length of about half a metre, found over much of eastern Australia including Tasmania.

**H. S. Curtis**
By email, no address supplied

# ❓ Back on track

*While working in the garden, I saw a beetle walk past, take a wrong step and land on its back. Without my intervention it would have stayed in this position and probably died. Why is it that millions of years of evolution have not eradicated this basic and potentially lethal design fault?*

**Greg Parker**
Brockenhurst, Hampshire, UK

If your correspondent had left the beetle in place on its back it probably would not have remained as it was until death.

Beetles and other insects have a variety of mechanisms which they can use for righting themselves in these circumstances which, as the writer presumes correctly, must arise often and hazardously.

The most famous mechanism is used by the click beetles (Elateridae), which are able to launch themselves into the air by the sudden release of a blunt spine which is kept under pressure in a specialised groove on the venter.

As many readers will have noticed, the click beetle often makes several attempts before it lands on its feet, but its success, given time, is assured.

Other less sophisticated beetle correcting mechanisms include spreading the wings, reaching out with the legs, and rocking the body in a forward-aft or side-to-side motion.

**Christopher Starr**
Department of Zoology
University of the West Indies
St Augustine, Trinidad and Tobago

Only a minority of beetles possess a body plan that poses such a problem. For example, I have worked with several species of ladybeetle (Coccinellidae) in the laboratory, and most are able to right themselves with relative ease.

The species that do find themselves stranded on their backs tend to be the larger varieties that possess strongly convex elytra (the first pair of hardened, protective wings).

Ladybeetles that do become stranded on a smooth surface will eventually unfold their membranous hind wings, which are normally hidden beneath the elytra, and then use these to right themselves. Part of the answer, then, is that very few species become stranded and those that do eventually flip themselves over by means of their hind wings.

Over the long course of evolution it was probably quite rare for beetles developing in temperate forests and grasslands to encounter totally smooth surfaces or bare soil that was devoid of plant litter. Under normal circumstances, grass blades, fallen leaves and plant stems would offer a convenient hold for beetles that happened to become overturned.

The reduced rate of predation and numerous other benefits that are conferred by a hard protective covering, which far outweighs the occasional stranding, has contributed to the enormous evolutionary success of beetles. In terms of both absolute numbers and numbers of species, beetles are the most successful group of animals on the planet.

**Tom Lowery**
Pest Management Research Centre
Ontario, Canada

I doubt whether the beetle was a healthy specimen that just happened to fall over and was unable to right itself. It is more likely that it was an old, sick or diseased specimen that was nearing the end of its life. When this happens in beetles, they lose a great deal of their mobility and coordination and they become very unstable when walking. They frequently fall over when placed on a hard flat surface and are unable to right themselves.

I have observed this countless times in a number of beetle groups. In fact, while growing up I lived near Milwaukee, Wisconsin, in the US. We had a fairly large population of *Carabus nemoralis*, which is a ground beetle that was introduced from Europe into the US. I would frequently find beetles on the sidewalks on their backs. No matter how many times they were righted, they would invariably end up on their backs again, soon to die. I also observed beetles stagger out of the vegetation bordering the sidewalk, only to fall onto their backs. If these beetles were placed on their feet, even in the vegetation, they would stagger about and would fall onto their backs again when they encountered the sidewalk.

So, I suspect that the poor design is really a combination of dying beetles coupled with a smooth, hard surface – one that is not normally found in nature. Considering that roughly one out of every five living creatures is a beetle, and that they occupy virtually every niche and habitat known, I would suggest that beetles are, in fact, very well designed animals.

**Drew Hildebrandt**
By email, no address supplied


# Myopic mammals

*I would estimate that about 40 per cent of people that I know need glasses or contact lenses for distance vision. Assuming that this sample is typical of the human race, I would like to know why it is that eye problems prevalent in humans such as myopia (short-sightedness) seem very rare in wild animals. As far as I know, myopia is a genetic condition and so is not usually acquired by habits such as reading small print (otherwise one would expect recovery after stopping the habit). Obviously, it is not easy to test the eyesight of an animal, but if the incidence of myopia is as high in wild animals as it is in humans then how can the animals survive?*

**Stephen White**
Surbiton, Surrey, UK


For most nonhuman mammals the ocular refraction (the lens power required to form a clear retinal image of an object at infinity) for optimal distance vision tends strongly towards emmetropia, or perfect vision. Similarly, the variation of distance refraction and the presence of astigmatism is lower than for humans. It should be noted, however, that the sample tested is far smaller for mammals than for humans.

These trends have been noted by my co-workers Clive Phillips, Jacob Sivak, Robin Best and Bill Muntz, and myself, in a number of different studies using standard clinical opto-metric techniques (obviously avoiding those that require a verbal response).

The studies have covered such disparate species as domestic sheep, guanaco (a llama), polar bear, manatee and three-toed sloth. Such findings would support the suggestion that a myopic mammal would be at a natural disadvantage.

**David Piggins**
Bangor, Gwynned, UK

There probably is a genetic factor in short-sightedness, but that does not explain why it is so common in modern society. Those who regularly focus their eyes over longer distances, such as sailors and mountaineers, are apparently less likely to become myopic. It seems likely that the muscles on either side of the eye can be trained to contract the eye, thus overcoming short-sightedness. Once a person starts wearing glasses, the need for such adjustment disappears.

**Brynjolfur Thorvardarson**
Southampton, Hampshire, UK

The fact that about 40 per cent of people you know wear glasses or contact lenses does not really indicate a malfunc-tion in the entire human race. If you travel among primitive peoples of the world, you will find numerous examples of keen sight that seem almost super-human to the Western mind.

**Kevin Wooding**
Oxford, UK

Myopia often has a genetic component, but this isn't the whole story. People who do close work are often myopic (tailors are the classic example) but in the past it was usually assumed that it was their myopia that attracted them to such jobs: the idea that myopia could be acquired seemed too far-fetched. However, several decades ago it was observed that university students with better than average grades (who presumably read more) tended to be myopic, as were laboratory animals that were raised in a confined environment.

**C. R. Cavonius**
University of Dortmund, Germany

Primates brought up in captivity do tend to become myopic. Myopia is caused by the axial length of the eye, but changes in corneal power also affect sight. Most change is likely to occur in the growth phase during the few years after birth, but may continue to a lesser extent for the next three decades in humans.

Better evidence comes from chicks, where 10 dioptres of myopia or hypermetropia can be induced by contact lenses, and reversed, over a few weeks. There is a dramatic change in the posterior segment of the eye, which is accounted for by alterations in the rate of growth of the eye as the chick ages. Just like humans who squint to improve their vision, chicks have the capacity for corneal and lenticular accommodation, but this appears to exert little influence upon growth.

Whether human myopia can be arrested or reversed is the subject of some debate in ophthalmology. There does seem to be something approaching an epidemic of myopia, especially in the Far East, which cannot be explained purely by genetic or occupation selection.

**Matt Cooper**
Brighton, Sussex, UK

Among older people, acquired defects of vision are more common than inherited myopia. At ages beyond those that would have been achieved in the wild, human eye tissues distort and lose flexibility, causing astigmatism and presbyopia. The lens may go opaque from cataracts, or it may darken so that it needs more light.

Domestic animals may show similar defects when they get beyond the life expectancy in the wild, and old dogs or cats often go blind from cataracts. Replacing a pet dog's lens is practically a routine operation nowadays. One need not give the dog glasses afterwards – just as long as it can see clearly enough to get at its food dish. It doesn't have to read the brand name on the tin.

**Jon Richfield**
Somerset West, South Africa

## ❓ Milking the issue

*Cows eat lush grass in summer but prepared dried feed in winter. Does the milk that I pour on my breakfast cereal differ in any way throughout the year?*

**Graeme Mawson**
Newcastle upon Tyne, UK

More than fifty years ago, I spent some time living on my aunt's croft in Sutherland.

In winter, her cow Bella was fed largely on turnips, and during those months Bella's milk also tasted of turnip. But this taste disappeared in summer when she was fed on grass.

Nowadays, dairy cows are fed on a fermented stored grass called silage during the winter and this is supplemented with processed feed containing animal protein.

The winter milk no longer smells or tastes of turnip, but I assume that if turnip used to contribute to the composition of milk then the modern winter feed probably does too. In retrospect, I think I would prefer the turnip.

**Ian Sutherland**
Birmingham, UK

Your correspondent probably assumes that the milk he pours onto his breakfast cereal every morning is much the same as the milk that comes out of a cow. In fact, most drinking milk is pasteurised and standardised for fat, so this quality parameter is invariable throughout the whole year. Milk may in future also be standardised for protein – in effect removing high-value solids from the milk for further processing.

As a dairy farmer drinking raw milk, both I and my family might detect slightly higher levels of fat and lactose at certain times of the year but, unless the cows graze on a patch of wild garlic, the taste of the milk remains the same.

I believe that in parts of France, certain cheeses are only made from milk produced in a particular season and from cows that graze on pastures with specific herbs.

Pasteurising milk probably removes taste by damaging natural enzymes. If anyone could achieve the antibacterial effects of pasteurisation without also damaging the taste, the process would be invaluable to the dairy industry and would also enable your correspondent to experience the taste of milk straight from the cow.

**Mark Pearse**
South Molton, North Devon, UK

The effect of different feeds on the content of milk is actually only slight. The nutrients affected are the fat-soluble vitamins A and E, folate and iodine. There is no significant difference in any of the macronutrients such as protein, carbohydrate and fat.

Vitamin A in whole milk varies from 69 micrograms per

100 millilitres in the summer to 44 micrograms in the winter. Whole milk is a useful source of vitamin A, particularly for young children, and this is one of the main reasons why milk is recommended for children under two years.

The variations in vitamin E and folate are both small: vitamin E averages 0.1 milligrams per 100 millilitres in summer and 0.07 milligrams in winter, while folate rises from 4 micrograms per 100 millilitres in summer to 7 micrograms in winter.

Iodine content averages 7 micrograms per 100 millilitres in summer and 38 micrograms in winter because cattle consume greater amounts of iodine-supplemented manufactured feed in the winter months.

The widespread addition of iodine to animal feed has meant that milk and dairy products are now major sources of iodine in the British diet – a factor which has helped significantly to eliminate goitre.

**Sarah Marshall**
National Dairy Council, London, UK

## ❓ In-flight meal

*During migration the ruby-throated hummingbird (Archilochus colbris) tanks up with a few drops of nectar for the last time on the northern shores of the Gulf of Mexico. It then flies non-stop for at least 800 kilometres to reach the shores of the southern Gulf. Can anyone calculate the metabolic fuel efficiency of these birds that fly so far on so little, and how does this compare to a human?*

**Martin Bradfield**
Lohhof, Germany

This question may involve some erroneous assumptions. Before departing for the Yucatán peninsula, a hummingbird spends weeks gorging on arthropods and does not merely consume 'a few drops of nectar'. It puts on enough fat to nearly double its weight: a female can grow from 3.2 grams to around 6 grams, and can barely get airborne. When, after anything up to 22 hours, it reaches its destination, it will weigh around 2.7 grams, having consumed the fat and often some muscle tissue. Many do not complete the trip.

The average metabolic rate for the black-chinned hummingbird is 29.1 ± 6.3 kilojoules per day. A man, metabolising energy at the same rate, would have to consume twice his weight in meat a day, or 45 kilograms of glucose, and his body temperature would rise to over 400 °C.

**Lanny Chambers**
St Louis, Missouri, US


The basal metabolic rate is the measurement of how much oxygen an organism uses when at rest. Just the fact that the hummingbird has a very high ratio of body mass to surface area gives it a basal metabolic rate that is 12 times as high as a pigeon's and 100 times that of an elephant.

The metabolism of the ruby-throated hummingbird is much lower when in torpor than in flight. To go from torpor to an active state takes it about an hour. The heart rate rises from 50 beats per minute to 500, and its temperature from 10 °C to 40 °C. When in full flight its heart pumps at 1,260 beats per minute and its wings beat 50 to 200 times per second. And this is still more energy-efficient than a human taking a walk.

Ruby-throated hummingbirds belong to a group of birds known as passerines, which have three toes facing forwards and one toe back. Passerines tend to have a metabolic rate that is as much as 70 per cent higher than either non-passerine birds or mammals. Their muscles are made up of about 35 per cent mitochondria with densely packed cristae – infoldings of their inner membrane – which makes them at least twice as efficient as human mitochondria.

To achieve the same efficiency, humans would have to have muscles composed of 70 per cent mitochondria. And even then the muscles could not work because there would be too few myofibrils in them.

While wintering in Mexico, the bird doubles its weight by building up considerable fat reserves to use on its long journey north. In fact, most ruby-throated hummingbirds travel along the edge of the Gulf of Mexico, eating a little along the way, but some do take the short cut from Florida to Yucatán using their fat reserves and catching gnats along the way.

A number of key factors make the ruby-throated hummingbird so efficient. Its pectoral muscles are red meat, which is rich in the oxygen-carrying protein myoglobin; the muscle has a high capillary-to-fibre ratio, giving it a good blood supply; it has an energy-rich diet of nectar that is stored as fat; it can eat nectar from any species of flower so it does not waste time looking around for a particular source, and will also eat any insects; its tongue is fringed so that the nectar is effortlessly drawn up by capillary action; and it operates at high temperatures to make its metabolic reactions more efficient.

It has been calculated that this species eats as much as three times its own weight in a day, during which it is awake for 16 hours on average. This is the equivalent of an 83-kilogram human eating 125 kilograms of hamburgers every day, or 1,335,000 kilojoules.

**Mike Ball**
Gorinchem, Netherlands

# ? Burnt offerings

*Almost all dog food contains ash as an ingredient. Sometimes*
*it makes up as much as 14 per cent by weight. I have always*
*thought of ash as being toxic waste, containing all sorts of noxious*
*elements, so why is it added to dog food and what type of ash is it?*

**William Davidson**
Strathaven, Lanarkshire, UK

You will be relieved to hear that ash is not added to pet foods.
It is a way of describing the mineral content of pet food. The
ash you see listed is part of the guaranteed nutrient analysis:
legally the pack must state how much of the food is protein,
fat, fibre, water and ash.

Ash is measured by heating the pet food to temperatures
of around 550 °C, and burning off all the organic components
to leave just the inorganic or mineral residue. If the mineral
content of pet food sounds high, it is important to remember
that our domestic carnivores were designed to eat carcasses
that are full of bones containing minerals, and a well-designed
pet food will reflect this in its composition.

**Kim Russell**
Registered pet nutritionist
North Molton, Devon, UK

This is a misreading of the label on the product. Ash is usually
given under 'typical analysis' or a similar heading, not under
the ingredients list.

Foods are often described in terms of their nutritional
content by carrying out a proximate analysis. This is done
because it is much quicker and cheaper than carrying out a
detailed analysis of the nutrients.

The protein content is determined by analysing the
nitrogen content, using a technique called the Kjeldahl

method, and multiplying by a conversion factor to obtain a crude protein figure.

The fat content is derived by gravimetric extraction using a suitable non-polar solvent (usually petroleum ether), while the water content is obtained by drying, and the mineral content is found by burning off all the organic material in a muffle furnace to obtain the ash. The carbohydrate content is often estimated by subtracting the former components from the total weight.

So ash is not added as an ingredient but is instead an indicator of mineral content. These minerals will be chiefly potassium and phosphorus with smaller amounts of calcium, iron, magnesium, sodium and zinc, and trace amounts of many others. Historically, manufacturers often boosted the mineral content of dog food with bone meal to raise calcium levels but, because of concerns about BSE, they now tend to use fish meal instead.

You might see ash levels of 14 per cent in a dry meal for dogs, but tinned products often have around half this level. The composition of the food will affect the ash content, but the elements are likely to be beneficial or neutral to the dog's health and not noxious or toxic at the concentrations in the product. It is worth pointing out that dogs should always have access to fresh water to ensure they can urinate away any excess potassium or sodium.

**Brian Ratcliffe**
Professor of Human Nutrition
The Robert Gordon University
Aberdeen, UK

# ? Floundering about

*How do certain animals, such as the flounder fish, change their colour to match their background? More specifically, if you made a tiny blindfold for the flounder, would it still be able to match its surroundings?*

**Nick Axworthy**
By email, no address supplied

Many fish in the teleost group, such as the minnow, change colour in response to the overall reflectivity of their background. Light reaching their retina from above is compared in the brain to that reflected from the background below.

The interpretation is transmitted to the skin pigment cells via adrenergic nerves, which control pigment movement. Teleost skin contains pigment cells of different colours: melanophores (black), erythrophores (red), xanthophores (yellow) and iridiophores (iridescent). Pigment granules disperse through the cell from the centre. The area covered by the pigment at any time determines that cell's contribution to the skin tone.

Many flatfish, including flounder, go further than overall reflectivity and develop skin patterns according to the light and dark divisions of their background. This seems to involve a more discriminating visual interpretation and produces distinct areas of skin with predominantly, but not exclusively, one type of pigment cell. For example, black patches contain mainly melanophores and light patches mainly iridiophores, which can even produce a chequerboard appearance if the fish is lying over a chequered surface.

Since these responses are visual, blindfolding the fish would result in all the components of the chromatic system being stimulated equally. The fish would adopt an intermediate dark or grey skin tone similar to that on a dark night.

Over time, hormonal responses via direct light stimulation of the pineal gland through the skull also affect the amount of pigment and number of cells, hence the 'black' plaice sometimes sold in the UK, which have come from the sea around the dark volcanic seabed off Iceland.

**Cliff Collis**
London, UK

Many animals change the shade or even colour of their skin in response to certain stimuli. In cephalopods such as the cuttlefish, pigment-filled sacs can become extended (flattened) by the action of radially arranged muscle fibres that are controlled by the nervous system. Colour change in these animals is both rapid and spectacular.

In crustaceans and many fish, amphibians and reptiles, specialised dermal pigment-storing cells called chromatophores relocate the pigment internally. The pigment in these chromatophores is either concentrated in the centre of the cell, or dispersed when the pigment fills the cell to the edge.

Imagine a white floor with a small pot of black paint standing in the middle. From above, the floor will look very light, despite a substantial amount of pigment seen as a small black spot in the middle. When the same paint is spread over the floor, the floor looks black. The beautiful trick of the black chromatophores (known as melanophores) is that they can reverse the process, concentrating the pigment in a small area.

Flatfish, such as plaice, flounder and others, are expert at imitating not only the general shade of the surface on which they rest, but also patterns of dark and light material. Not surprisingly, perhaps, their eyes are used to perceive the shade and patterns. Light hitting the retina from above affects the ventral or lower area of the retina, while light reflected from the bottom strikes the dorsal or upper retinal surface.

If the light intensities from the two areas are similar, a

signal causes the pigment of the melanophores to be concentrated in the centre of the cell, so the fish turns pale. On the other hand, when the bottom is dark the two areas of the retina receive very different light intensities, and the reverse of the signal causes pigment dispersion and a dark fish. The masters of disguise, the flatfish, can also discern patterns in the bottom surface and imitate them by regulating nerve activity to groups of melanophores.

**Stefan Nilsson**
Gothenburg University, Sweden

# ❓ Biosphere

*Hypothetically (because otherwise my mum would get mad), if I were to put my brother in a perfectly sealed room, how much plant life would I need in that room in order to maintain a balance of oxygen and carbon dioxide such that both my brother and my beloved plants may continue to live?*

**Gene Han**
Iowa, US

To simplify matters, you could supply his meal through an airtight hatch. The plants would then only need to provide his oxygen. If he spent all his time eating and dozing, he would need about 350 litres of oxygen per day (the amount of oxygen in 1.7 cubic metres of air). This much oxygen is produced in full sunlight by typical vegetation covering a floor area of between 5 and 20 square metres. Using the most productive 'C4 plants' such as sugar cane, you could reduce the area needed to 2.5 square metres. Your brother would exhale 350 litres of carbon dioxide per day, which would enable the plants to grow with an increase of dry weight of 430 grams per day.

Now let's muddy the waters. If his windows plus artificial lights supply 10 per cent of full sunlight, multiply the required area of greenery by a factor of 10. If the lights go out at night, double the area – more in winter. Plants photosynthesise during the day more rapidly than they respire at night. Therefore, as a reasonable approximation, you can neglect the extra oxygen that plants consume at night.

If you don't intend to feed your brother, but hope he will survive by eating the plants, remember that most material a plant synthesises is indigestible, so double the area again. The inedible parts of the plants plus your brother's faeces would need to be decomposed or burnt to carbon dioxide to recycle the carbon they contain. So, if your brother is a well-trained plant physiologist, this ambitious biosphere might need to be a plant-filled room about 20 metres square.

Here is the basis of my calculations:

The daily energy requirement of an adult dozing is 1,750 kilocalories per day. The energy content of 100 grams of sucrose is 400 kilocalories. Therefore an adult needs 1,750/4 = 438 grams of sucrose per day = 1.28 moles sucrose per day.

Respiration of this requires $1.28 \times 12 = 15.36$ moles of oxygen per day. One mole occupies 22.4 litres, so this corresponds to $15.36 \times 22.4 = 344$ litres of pure oxygen per day.

Photosynthesis rates of plants in the field under optimal lighting are between 10 and 30 micromoles (up to 70 in C4 plants) of carbon dioxide fixed per square metre per second (0.86 to 6.05 moles per square metre per day). For each mole of carbon dioxide the plants fix, they liberate a mole of oxygen.

Therefore the area required is somewhere between 18 square metres for the less productive plants down to 2.5 square metres for C4 plants.

**Stephen Fry**
Institute of Cell and Molecular Biology
Edinburgh University, UK

# ☑ Blood brothers

*At the risk of flogging a dead, er, penguin. Why don't polar bears'*
*feet freeze?*

**Paul Newcombe**
Zurich, Switzerland

Unlike the penguin with its fancy internal plumbing, the reason that polar bears' feet do not freeze is good insulation, pure and simple.

Polar bears (*Ursus maritimus*) are just about the best-insulated animals on the planet, certainly among those species of mammal that do not live primarily immersed in water. An adult bear has 10 centimetres of blubber beneath its skin, which in turn is covered by a thick coat of fur. This fur relies not only on its density, but also on its unique structure: each hair is a hollow tube, so that air is trapped inside the hairs as well as between them. Even without covering its nose with its paws (as it is reputed to do, although the evidence is very limited) a polar bear is almost invisible to heat-sensitive infrared photography or the latest military image-intensification technology.

The polar bear also has very hairy pads on its feet, and the tough skin is extremely callused on the underside of the paws, so there is a sturdy layer of dead tissue between the ice and any blood vessels.

There may also be another factor at work. The underside of a polar bear's paw is dotted with dozens of papillae – small nipple-shaped extrusions of even more callused skin – which provide extra grip in the same way as the studs on a footballer's boot. It is these papillae that enable a polar bear to accelerate to a very respectable pace on the ice and overcome its awesome inertia. They also prevent it skating out of control, past a potential meal.

On really compacted ice, the bears tend to lift part of the underside of the paw clear of the surface. The papillae enable an additional cushion of insulating air to be trapped between most of the pad and the ice.

Such highly developed thermal adaptations can, however, be a double-edged sword. A bear attempting a brisk trot in ambient temperatures of 10 °C or greater would succumb, almost immediately, to a fatal attack of heat stroke. During the Arctic summer it can often be far hotter than that, limiting the polar bear's ability to function as a hunter.

This potential cramping of the polar bear's style may prove as fatal to the species' chances of survival as the actual destruction of its territory. If global warming causes the polar bear to die out, it would surely be the most terrible irony that this was because it had mastered the art of conserving the very energy that a profligate humanity has squandered so obscenely.

**Hadrian Jeffs**
Norwich, Norfolk, UK

# ❓ Carriers of death

*Do mosquitoes get malaria? Do rats catch bubonic plague? If not, why not?*

**Year 5, Christopher Hatton School**
London, UK

Congratulations to the children for asking such a penetrating question.

Rats can get quite sick from plague fleas and some will die, but usually not too quickly. Plague-carrying rats are at their most dangerous when they are about to die, because their fleas leave them as soon as they are dead to find new hosts.

The malaria parasite *Plasmodium* does not usually kill its host mosquitoes, though it may take a high enough toll that it is better for the mosquitoes not to get infected.

If we could breed mosquitoes that were resistant to the parasite we might find that they outcompete ordinary mosquitoes, and this might ultimately help get rid of malaria. This kind of strategy would not work with yellow fever, as the mosquitoes that carry the virus responsible for the illness in humans hardly seem to be affected.

**Jon Richfield**
Somerset West, South Africa

*Plasmodium*, the parasite that causes malaria in humans, infects mosquitoes. The mosquitoes then transmit it to people when feeding on their blood. As for the plague microbe, *Yersinia*, it blocks the gut of the flea that transmits it. As a result, when an infected flea feeds on the blood of a human or rat, it will regurgitate some blood containing the microbe and so spread the germ to a new host.

To address the question directly, the important thing to note is that being infected with a microbe or other parasite does not necessarily cause disease, because it is often in the interest of the microbe to cause no harm to its hosts.

However, the mosquitoes that transmit *Plasmodium* are affected by it, as the parasite grows in their salivary glands. Such infection can reduce the ability of the salivary glands to function and thus the viability of the mosquitoes.

A related parasite called *Theileria*, transmitted between cattle by ticks, can damage the gut and salivary glands of the ticks, and can even kill them in the laboratory. Epidemiologists make a point of studying the extent of such effects under natural conditions.

**Alan R. Walker**
Edinburgh, UK

# ▢ Topsy-turvy world

*Why don't bats get dizzy when they hang upside down? Or do they?*

**Year 5, Christopher Hatton School**
London, UK

Dizziness is a sensation humans describe when they feel a sense of motion, even when not moving. It can be associated with queasiness or nausea, and sometimes vomiting. Other types of dizziness include motion sickness and vertigo, which often manifests itself as a spinning feeling, or other sensations such as light-headedness or heavy-headedness.

It is impossible to know for sure whether or not an animal is dizzy, because it cannot communicate such feelings. However, it is possible to infer an animal is dizzy from how it behaves. For example, if an animal is aimlessly walking in circles, it is probably dizzy.

Motion sickness occurs when there is excessive stimulation of the inner ear or from a conflict between sensory information from different sources, such as from the inner ear and the eyes. The balance mechanism of the inner ear is complicated, and includes sensors that detect both movement and orientation with respect to gravity, even when an individual is not moving. Bats have such a balance mechanism, and in addition use echolocation.

The parts of the inner ear that are important for orientation with respect to gravity are called the otolith organs: the utricle and the saccule. It is these parts of the inner ear that would be activated while the bat was hanging upside down. Stimulating these parts of the inner ear, however, would not necessarily lead to dizziness, especially in a dark cave where there is no conflict between information from the inner-ear balance mechanism and vision.

The bottom line is that bats are used to hanging upside down without showing any behavioural changes that would suggest dizziness or motion sickness. But because we cannot ask a bat directly whether or not it is dizzy, we cannot be certain about the effects of hanging upside down.

**Joe Furman**
Editor of the *Journal of Vestibular Research*
University of Pittsburgh
Pennsylvania, US

When you think of bats, you usually think of them in one of two conditions: hanging upside down resting, or flitting about pulling high-g turns in the dark. So why don't they get dizzy?

Bats have evolved a number of adaptations to allow them to hunt and hang without the problems that humans would face. First, some bats have specialisations in the vestibular portion of their inner ears – the portion that generates sensory signals for controlling balance. Their sacculus, which in humans acts as a gravity sensor to help us stand upright, is slightly rotated forwards. This enables it to act more as a pitch detector, which is more useful in flight. Second, their semicircular canals, which sense rotation of the head, have an internal structure more like a bird's than a human's. This probably allows them to make high-speed turns without the fluid in the canals sloshing back and forth too much. Lastly, if you photograph bats in flight with a high-speed camera, you notice that they keep their heads very stable except in the most violent turns.

But it is the way in which bats sense the world that probably gives them immunity to dizziness. All the vestibular system does is tell you about changes in acceleration of your head. It requires other senses to pin down your position and motion in the outside world. We primarily use vision to do this, but vision is very slow. Anything you look at that takes a second or less to cross 30 degrees of your vision appears smeared. Echolocating bats, while not blind, rely more on biosonar, an especially precise form of hearing that lets them build up 3D images from echoes.

Echolocating bats emit brief sonar chirps from 30 to more than 150 times per second, and respond to changes in echoes of less than a microsecond. These bats integrate echolocation with their vestibular system, so they are working with a faster, more precise positioning system than humans do with vision. Because dizziness and motion sickness usually arise when signals from the vestibular system conflict with those from other sensing systems, bats are less likely to show motion sickness than other mammals.

**Seth Horowitz**
Assistant professor of neuroscience
Brown University, Rhode Island, US

# 5 Our planet, our solar system

## ❓ Landlubbers

*Where on our planet is the furthest point from any sea? I'm hoping it is the middle of Asia somewhere because I'm travelling that way soon, and want to stand at that point in swimming trunks, snorkel and mask.*

**Hugh Jones**
Slapton, Northamptonshire, UK

The furthest point from the sea or, to give its technical name, the continental pole of inaccessibility (CPI), does lie in Asia. It is located at 46° 17′ N, 86° 40′ E, in the Dzoosotoyn Elisen in Xinjiang, China, and is 2,648 kilometres from the nearest coastline, at Tianjin on the Yellow Sea. Although its location was calculated long ago, it wasn't visited by surveyor-explorers until 27 June 1986, when it was reached by British cousins Nicholas Crane and Richard Crane.

The Cranes travelled there by bicycle, crossing the Hindu Kush and Gobi deserts, to raise funds for the Intermediate Technology Development Group (which has since been renamed Practical Action) – a charity that supports technological advances in developing countries.

Twenty years before that, however, the CPI attracted the attention of another group with very different interests. Its unique geographical status gave it considerable significance among western nuclear strategists debating the relative merits of weapons systems. For proponents of the submarine-launched Polaris missile, the ability to hit any point on the Earth – even if there is nothing there worth hitting – became a key point in the public relations battle with the sponsors of land-based and air-launched weapons.

When the A3 version of Polaris brought the CPI within range in the late 1960s, it was hailed as a technological triumph – particularly by the UK's Ministry of Defence. They did not, however, trumpet the fact that to strike the pole a large nuclear-powered submarine would practically have to visit Tianjin docks.

Ironically, by the time the western navies acquired the capacity to bombard all of China with submarine-launched missiles, the region around the CPI was probably featuring more prominently on the targeting lists of generals in Moscow, rather than London or Washington DC, as Xinjiang acquired vital strategic significance in the Sino-Russian confrontation of the last quarter of the 20th century.

Finally, while appreciating the irony in your correspondent's choice of attire, he should reflect that the CPI is subject to extreme climatic continentality: summers are hotter and winters are colder than many places of similar latitude because it is so far from the moderating influence of the ocean. 'Elisen' means 'desert' in the local Chinese Uighur dialect, and although the location is certainly sandy, it is no beach.

Indeed, this part of Xinjiang might be considered an extension of the Gobi, which is a decidedly cold desert where temperatures drop to -40 °C in winter. At the other extreme, during daytime in summer it can reach a blistering 50 °C, though temperatures can vary by as much as 32 °C within a 24-hour period. If he insists on wearing his trunks, I suggest that he keeps a warm pullover handy, just in case.

**Hadrian Jeffs**
Norwich, Norfolk, UK

# ❓ Fooled in Blackpool?

*From the top of Blackpool Tower (approximately 150 metres) on the UK's west coast, can you see the curvature of Earth along the Irish Sea horizon? I thought I could, but my friend disagreed. If I'm wrong, how high would we have needed to be?*

**Mark Ford**
Bolton, Lancashire, UK

While camped at 6,000 metres altitude in Peru in 1962, I and some colleagues asked ourselves this question about the Pacific horizon. We actually only saw the curvature by comparing the horizon (about 277 kilometres away) with a nylon thread stretched tight and level between two ice axes.

While standing atop Blackpool Tower, if you sight the seaward horizon over a level, 1-metre straight edge, which is held 1 metre in front of you, trigonometry shows that the ideal horizon would appear to be almost a millimetre higher at the centre of the straight edge than at the ends. This is a much smaller effect than typical atmospheric distortion which, in effect, means there is no visible curvature.

From our camp in Peru, the difference was almost 6 millimetres – easily visible when compared with a straight edge. Even so, the curvature was not apparent when simply looking at the horizon.

**Charles Sawyer**
Byron Bay, New South Wales, Australia

As the radius of the Earth is 6,373 kilometres, a little trigonometry tells us that if you are at the top of a tower of height h metres, the horizon will be at a distance of approximately $(2 \times 6373 \times h)^{1/2}$ kilometres.

For a tower 150 metres high, the horizon will be 44 kilometres away and displaced downwards from a true horizontal line by about 0.39 degrees. If you hold a 1-metre stick horizontally 1 metre in front of you, seemingly touching the horizon at the midpoint of the stick, the ends will appear to be 0.8 millimetres above the horizon. That's pretty hard to see with the naked eye.

**Eric Kvaalen**
La Courneuve, France

When out in the mid-ocean, up at the top of the main mast, the horizon is a horizontal line right round the field of view. The higher the mast, the lower the horizon appears to be, but it is still a horizontal line.

**John Eagle**
Wilmslow, Cheshire, UK

The short answer to this question is that the curvature is not obviously visible from anywhere on the Earth's surface. Pilots of Lockheed U-2 and SR-71 Blackbird aircraft suggest that the Earth's curvature only becomes clear at an altitude of about 18 kilometres. Indeed, it has been photographed from Concorde cruising at this altitude. The curvature can be inferred at sea level, though. For example, ships disappear over the horizon from the bottom upwards, as if sinking into the sea.

**Mike Follows**
Willenhall, West Midlands, UK

Seeing the curvature of the Earth can mean either seeing the surface of the Earth in front of you fall away towards the horizon, in the same way that you see the ground fall away when standing on a rounded mountain top, or seeing the horizon as a curved rather than horizontal line.

It is actually possible to see the curvature of the Earth, in the second sense outlined above, at any height: for example, sitting on the beach, standing on the deck of a ship or looking out of a plane window. This is to be expected, because a view from any point on a sphere such as the Earth will give the horizon as a disc. The height of the viewpoint will simply determine its size.

The visual cues employed to see the curvature of the Earth are many, but judging the line of the horizon relative to the horizontal is generally not one of them. Instead, two more obvious cues are noting that the distance of the horizon is the same in any direction, and seeing that the texture gradient – the way a view changes in appearance and perspective with distance – of the sea or land is constant within that distance.

I agree that increased viewpoint height will yield a richer set of cues, especially those associated with seeing the horizon in the second sense, delivering a more obvious curvature. Nonetheless, this curvature can still be noted at sea level.

**John Campion**
Psychologist and vision scientist
Liphook, Hampshire, UK

Every time I read yet another theoretical contribution to your debate over whether the curvature of the Earth is apparent from the top of Blackpool Tower, I turn to the front cover of the magazine just to check that the word 'scientist' is really still there.

I would have thought that by now someone would have followed the scientific method: just do the experiment and report the result. I would certainly have had a look if I lived a little closer.

**John Twin**
Ross-on-Wye, Herefordshire, UK

# ❓ Put that light out

*If the sun was extinguished or there was a permanent worldwide
eclipse, how long would it take for us all to freeze to death, and
what could we do to try to avoid it?*

**Darren Darby**
London, UK

A back-of-the-envelope calculation suggests the whole Earth
might freeze solid within 45 days, radiating away its thermal
energy according to the Stefan-Boltzmann law, which relates
energy loss of a body to its temperature. My calculations
assume the vast bulk of the Earth's captured solar energy is
stored in the oceans, which have an average temperature of
15 °C down to a depth of 35 metres. Energy carried by water
at greater depths doesn't count because it would rapidly
become isolated from the surface by ice floes.

With its smaller heat capacity, the land would freeze much
more quickly than the oceans. Air over relatively warm oceans
would rise, pulling in cold air from the continents. This would
chill the surface waters and might increase the circulation of
water, exposing it to the chilling, perpetual night.

Interestingly, the volcanic dust thrown up by the eruption
of the Tambora volcano in 1815 acted both for and against
cooling. The dust blocked out the sun, but it also reduced the
escape of thermal radiation from the Earth by dint of a greenhouse effect. Sunlight dimmed by 25 per cent for a while,
leading to a dip in global temperatures of 0.7 °C in 1816. But
the fall in temperature was small despite a big reduction in
sunlight, suggesting that the Earth might take longer to freeze
than 45 days. Indeed, freezing may well be delayed further
by the natural greenhouse effect that comes with our atmosphere and the thermal inertia of our oceans.

Nevertheless, Earth would still be able to support a colony

of humans. There would still be plenty of energy in the form of fossil and nuclear fuels, and geothermal heat mines. But without plants to replenish our oxygen supply, it would quickly run out, so we would need to build biospheres with artificial light for plant photosynthesis.

Thankfully, switching off the sun is an experiment the Earth will not undergo for another 5 billion years.

**Mike Follows**
Willenhall, West Midlands, UK

*It would be interesting to discover what effect on global cooling or warming the 2010 Icelandic volcanic eruptions have had – Ed.*

You can estimate how long it would take us to freeze by extrapolating from the rate of cooling that happens overnight. In areas with clear skies, the temperature can drop to freezing in less than 12 hours. In places with heavy cloud cover, water vapour traps infrared radiation before slowing down the rate at which it radiates away into space, so cooling takes much longer. Here, the temperature drops by perhaps 5 °C in 12 hours. However, without the thermal energy of the sun constantly evaporating water, this insulating cloud cover would quickly disappear. It is likely that most parts of the Earth's surface would be frozen within a few days. The only exception would be near the coastline, where it might take a few weeks because of the amount of heat stored in the oceans.

Could we stop this? Perhaps we could quickly burn the world's forests to release large amounts of carbon dioxide to help trap infrared thermal radiation. However, with only fossil fuels, nuclear and geothermal energy left to rely on, we'd still freeze quickly. And if we didn't, we'd soon run out of food and oxygen.

**Simon Iveson**
Department of Chemical Engineering
University of Newcastle, New South Wales, Australia

The sun can't be turned off like a light bulb. It glows because its surface is about 5,500 °C and is heated by the nuclear fusion inside its core which is even hotter – about 15,000,000 °C. Even if fusion in the core could be switched off suddenly, the sun would continue to radiate light just as the heating element on an electric stove gives off heat for a time after you switch it off.

Obviously, the sun is bigger and hotter than a stove, so would continue to radiate heat and light for a long time. In addition, the energy produced in the core of the sun takes time to work its way out – millions of years if you track the energy by following the paths of individual photons. The sun would cool a tiny bit each year, but as the sun cools it would contract, releasing gravitational energy that would heat it and offset some of the cooling. That's how white dwarf stars continue shining. Suffice it to say that it would take many millions of years before our descendants even noticed.

However, if the sun suddenly vanished, the Earth would cool quickly. Unprotected people would start freezing in days, but they could survive much longer if they went down into deep mines, warmed by the Earth's heat.

**Jeff Hecht**
*New Scientist* contributing editor
Auburndale, Massachusetts, US

Even if the sun went out, ecosystems around hydrothermal vents along the Earth's mid-ocean ridges would continue to chemosynthesise using geothermal energy for a few thousand million years. So, business as usual for tube worms.

**Allan Mann**
Alnwick, Northumberland, UK

The previous correspondent was wrong to claim that it would be business as usual for tube worms if the sun went out. It is a modern myth that communities of chemosynthesising organisms at hydrothermal vents live in self-sufficient isolation. They don't. Most use the oxygen found in seawater for their metabolism. That comes, of course, from photosynthesis powered by sunlight in the near-surface waters and on land.

**Mike Cotterill**
Freshwater, Isle of Wight, UK

# ❓ Water world

*From reading your books I now know what percentage of the UK's surface area is roads, but having just returned from the Netherlands I would like to know what percentage of the surface area of that country is water.*

**Byron Hambleton**
Lille, France

Official statistics say that 19.3 per cent of the Netherlands' surface area is water. But any area of water less than 6 metres across is counted as land. So the total area of water will be well over 20 per cent.

**K. A. H. W. Leenders**
Historical geographer
The Hague, Netherlands

Ah – the Netherlands. The land of dykes, canals and windmills – and 18.41 per cent water. The Netherlands is the world's fourth most watery nation, behind the Bahamas (27.76 per cent), Guinea-Bissau (22.48 per cent) and Malawi (20.49 per cent).

The total area of the Netherlands is 41,526 square kilometres, of which 33,883 square kilometres is land – 27 per cent of it below sea level – and 7,643 square kilometres water. Despite the best efforts of nature, the amount of land is increasing, thanks to modern versions of the dykes and windmills for which the country is famous: since the 13th century, 10 per cent of the total land area of the country has been reclaimed from the sea as polders. These are beds of artificial lakes that are bounded by dykes, pumped dry by windmills and drained by canals.

However, it's not just the sea that the besieged Dutch battle. The Netherlands lies at the mouth of three major rivers: the Rhine, the Meuse and the Scheldt. Apart from the famous dykes protecting half the land from sleeping with the fishes, other dams and levees along these rivers protect against freshwater flooding.

**James England**
Woodville South, South Australia

# ? On a high

*A number of athletics and cycling world records have been set at high-altitude venues, for example during the 1968 Olympic Games in Mexico City. Presumably the air is thinner so there is less resistance, enabling them to run or cycle faster. But surely oxygen uptake at altitude is more difficult, so there must be a point at which altitude no longer favours athletes. What is this point and why? And which tracks or velodromes come nearest to it?*

**Carlos Loeb**
Madrid, Spain

Mexico City is situated at about 2,250 metres above sea level so, as the questioner correctly points out, air is less dense here because of the reduced atmospheric pressure – 580 millimetres of mercury (mmHg) compared with 760 mmHg at sea level – so you can't make a decent cup of tea because the water boils at 92 °C as opposed to 100 °C.

This reduced density of the air undoubtedly reduces the work that needs to be done by a cyclist, who must cut a path through it, but it poses problems in terms of oxygen availability. Although oxygen always makes up 21 per cent of the atmosphere, the fractional pressure it exerts in Mexico City is only 120 mmHg, compared with 160 mmHg at sea level. This means there is a noticeable reduction in oxygen pressure at the interface between air and blood in the lungs' alveoli, leading to mild hypoxia and a reduction in the amount of oxygen delivered to body tissue.

In the short term, athletes will breathe harder, leading to respiratory alkalaemia (increased blood pH), and heart output will increase to circulate the blood faster in an attempt to compensate. After several weeks' acclimatisation, the density of blood vessels in muscle and the number of red blood cells both increase, carrying more oxygen to the working muscle. In addition, the kidneys excrete extra bicarbonate to compensate for the alkalaemia caused by rapid respiration and consequent reduction in carbon dioxide levels. These effects will not confer any advantages to the exerciser while they remain at altitude, though they would briefly be of benefit if the athlete were to return to sea level.

At yet higher altitudes, above about 2,500 metres, heart rate and oxygen delivery are stretched to the utmost, beyond the ability of the body to compensate for them, and work output and athletic performance decrease.

It would appear that reduced air density is the only explanation for the performances at the 1968 Olympics, though doubtless athletes undertook extensive altitude training.

**Ian Jeffcoate**
Department of Veterinary Cell Sciences
University of Glasgow, UK

Altitude affects both running and cycling in two opposing ways. The power needed to overcome air resistance varies approximately with the velocity cubed and in direct proportion to the density of the air. The important consequences of this are that air resistance is far more significant at high speeds, and that it can be reduced by going to altitude, where the air is less dense.

The other effect of thin air is that the athlete receives less oxygen. In a race lasting less than 20 seconds, most of the energy comes from oxygen-independent glycolysis, in which the muscles break down carbohydrates without requiring large amounts of oxygen. This, combined with the sprinters' high speed, means that sprint race times will be quicker at high altitude.

Competitors in longer events depend more on aerobic respiration, so for any running race taking more than about a minute the benefits of altitude are lost. In cycling, the equation differs, because the higher velocity means that up to 90 per cent of the energy expended by a cyclist is used to counter air resistance, so almost all world record times would be faster at altitude.

The 400 metres athletics event sits somewhere between a flat-out sprint and an aerobic distance race. A 1991 study in the *Journal of Applied Physiology* suggests that the ideal altitude for this event would be between 2,400 and 2,500 metres, close to that of Mexico City. Indeed, one of the more enduring records set at the 1968 Olympics was achieved by Lee Evans in the 400 metres.

**Sam Baylis, Malvern**
Worcestershire, UK

# ? Moonbeams

*Why does the Moon appear as bright as a cloud in the midday sky, when it is a very dark body with an albedo of 0.07? Albedo is the ratio of the intensity of light reflected from an object to that of the light it receives from the Sun, and the albedo of clouds is around 0.6 to 0.8. This does not seem to apply to the difference between the brightness of the Moon compared with the brightness of clouds.*

**Nigel Scott**
Altrincham, Cheshire, UK

The albedo of the darker areas on the Moon is indeed around 0.07, but that of the mountains, rayed craters and cratered highlands is considerably higher (between 0.10 and 0.15). As viewed in a bright blue sky through binoculars or a small telescope, the dark areas appear almost indistinguishable from the surrounding sky.

**Mike Dworetsky**
Department of Physics and Astronomy
University College London, UK

The light we see reflected from the Moon is just that, reflected light. But we see the clouds by transmitted light because we are below the clouds and sunlight is passing through the clouds, not reflecting from them. Only high-flying jet travellers see light reflected from the tops of clouds and, even through the tinted aircraft windows, the clouds are 10 times as bright as the Moon, as the albedo figures suggest.

**Hazel Beneke**
Gatton, Queensland, Australia

Clouds have to be quite thick to reflect most of the sunlight.

Unfortunately, when they are sufficiently thick, they usually cover the whole sky and their bright tops cannot be seen. The converse is also true: when the sky is only partially covered with clouds, they are usually thin and no brighter than the Moon. However, if it happens that there are well-developed but sparse cumuli in the sky, and the Sun and Moon are on the opposite sides of the sky, then it can be observed that the top of a large cumulus is about ten times as bright as the Moon, as expected from their albedo ratio.

**Leszek Fraskinski**
J. J. Thomson Physical Laboratory
University of Reading
Berkshire, UK

## ? Sloping off

*In the hills south of Rome there is an area which, in the distant past, was volcanic. On the drive from Rocca di Papa to Albano there is a well-known gentle slope in the road which has an extraordinary property. If you stop your car, put it into neutral and then slowly release the brakes, the car will gradually, but perceptibly roll up the hill. I have observed this phenomenon twice, once as a passenger and once as the driver, both in broad daylight. Is there any physical explanation?*

**Nicholas Hutton**
London, UK

The same effect occurs on the A719 road in Ayrshire, UK at the Electric Brae. It has special warning signs because of the likelihood of meeting cars coasting uphill backwards, as baffled drivers are confused by their senses. Coming round the shoulder of a hill to enter a small, steep valley, the driver sees the road apparently falling towards the stream. But the road is traversing the side of the valley, rather than crossing the stream at right angles and it is actually slightly uphill.

**Douglas Stewart**
Department of Engineeering
University of Aberdeen, UK

Near Neepawa in Manitoba, Canada, there is a road called Magnetic Hill. You drive down a long gentle slope, stop, then release the brake and your car moves backwards, apparently up the hill. The local residents have finally had to admit that the hill is not really magnetic, but is actually a very convincing optical illusion produced by the local topography. No doubt the same applies on the road from Rocca di Papa to Albano.

**Peter Brooks**
Bristol, UK

The simplest explanation is an optical illusion, similar to one experienced by a friend and myself while we were on a cycling tour of northern Portugal. We were completely flummoxed to find ourselves having to pedal hard to make progress along a gently sloping, yet clearly downhill, stretch of road. The situation became positively surreal when a local came towards us also on a bike and apparently uphill, but with his feet resting on the handlebars, freewheeling.

**John Jeffries**
Potters Bar, Hertfordshire, UK

I have had the same experience twice in the past. I was driving across western Germany a number of years ago in an old Volkswagen Beetle. The autobahn along which I was travelling was, for many kilometres, a series of hills like a shallow sawtooth. My car could only manage about 100 kilometres per hour on the level and the prospect of driving uphill on this autobahn for 5 to 10 minutes at a time clearly exposed the power limitations of the engine. The uphill drive was foot-to-the-floor stuff the whole way.

At a certain point I became aware that I had lost the sense of whether I was travelling uphill or downhill, because while my eyes told me I was going uphill, the engine revs and the car's speed (which had reached 110 km/h) told me I was definitely going downhill.

I later had the same experience in Cornwall. The roads there are almost exclusively up and down, and at times it is impossible to tell whether your car is level or not.

**Dick Cullup**
Abu Dhabi, United Arab Emirates


I have experienced the same phenomenon at a similar site in Israel, near Jerusalem, exactly as described. There, the road is cut into the side of the hill, so that there is a drop on one side and a steep, rocky embankment on the other. The fact that the car appears to roll uphill is due to an optical illusion caused by the relative alignment of the embankment and the road. It is, of course, rolling downhill.

**Jonine Cortens**
Swansea, UK


I recall seeing a report of a similar occurrence in Australia many years ago. The road was near the landmark Hanging Rock, which was the setting for a mystery novel and a more famous film. Given its location, people were, I think, quite content to enjoy this apparent aberration and assume, or hope for, some unexplained, attractive force at work. The explanation provided was more mundane, but still curious. After careful measurement, the road was shown to be sloped, slightly, in the opposite direction to which it seemed.

**Mark Seto**
St Lucia, Queensland, Australia

A similar phenomenon can be experienced when driving across the centre of the island of Cheju Do off the southern coast of South Korea. This is a volcanic island with the remains of craters and eroded plugs and the experience is just as the questioner describes.

**Brian Smith**
East Molesey, Surrey, UK

This is almost certainly an optical illusion, like a similar road at Spook Hill, Lake Wales, Florida. This road seems to slope downwards for a short distance before climbing a hill, yet cars left in neutral roll backwards, apparently up and out of the depression. The mystery disappeared when an investigator used a spirit level to show that the dip is illusory, and the road is in fact uphill all the way. These observations were published in the autumn 1991 issue of *Skeptical Inquirer*.

**Jeremy Henty**
Cambridge, UK

The optical illusion that makes a downhill slope appear to be uphill is quite common in mountainous districts. It occurs when the true horizon is obscured by the surrounding hills and more distant hills in the direction of travel are rather lower than the nearer ones, giving one the impression of approaching the summit of a pass. This effect is enhanced when the valley floor approximates to a plane surface but is slightly tilted downwards in the direction of travel.

The fact that the questioner only observed this phenomenon in daylight tends to confirm the above explanation, as the surrounding hills would not be so obvious at night, even under a full moon. A further visit accompanied by surveying equipment would convince even the sceptical.

**Richard Burrows**
Tunbridge Wells, Kent, UK

I remember from a university surveying course the case of a water main in the Republic of Ireland where water was supposed to travel by means of gravity over a long distance. The drop was carefully calculated but very small. When water was introduced to the completed main, the engineers were horrified to see it flow uphill, back towards its source.

The reason turned out to be a local gravity anomaly which meant that gravity in the region acted not perpendicular to the surface of the geoid but at a slight angle backwards towards the water source. The very small anomaly was enough to overcome the even smaller gradient on the water main. Therefore the water really did flow uphill. I assume a small pump solved the problem.

**Oliver Moffatt**
Kendal, Cumbria, UK

I have experienced an occasion where a major river appears to run uphill. In the USA, Route 128 leaves Moab, Utah, to the east and runs alongside the Colorado river for several miles. At a certain stretch, the river appears distinctly to be flowing uphill – a highly disorienting impression. The rocks of the canyon in which the river flows are stratified and I suspect that it is the line of their bedding which plays the trick on the eye.

**David Cope**
Cambridge, UK

Like David Cope, I once saw a river 'running uphill'. It was on a flooded piece of unasphalted track near my school, in

Bradford, West Yorkshire. A beck had burst its banks in heavy rain and was running along the track. My class was out on the dreaded weekly cross-country run and we all stopped to comment on it. Although we were wet and tired at the time, our teacher brought us back the following week in warmer weather with string and spirit levels to prove that the track really did run downhill. Proof, perhaps, that innovative teaching really does have an effect? I remember his demonstration 30 years later.

**Derek Hite**
Manchester, UK

I remember seeing a related phenomenon while on holiday in Cornwall in 1978. But rather than a hill you could roll up, this was, apparently, a sloping lake. The gradient appeared quite noticeable, although I did not try unpowered waterskiing on it.

From what I remember, the lake was somewhere near Cape Cornwall, although I have failed to find it again on subsequent visits to the area. Can any locals let me know where it is?

**Chris Quinn**
Widnes, Cheshire, UK

It would be impossible to prove or disprove the antigravity phenomenon using a spirit level as described earlier. If the antigravity effect can act on a motor car, it can probably also act on a few millilitres of fluid containing a bubble, causing the bubble to float downhill instead of uphill.

**A. Stapleton**
High Wycombe, Buckinghamshire, UK

## ❓ One small footprint?

*Why can't one of our space telescopes, capable of seeing galaxies many light years away, be pointed at the site of the moon landings where one can assume there are some remnants from the visits? Would this definitively prove to any sceptics that humans landed on the moon?*

**Liza Brooks**
Shrivenham, Wiltshire, UK

The resolving power of a telescope – the size of the smallest object it can see at a given distance – is inversely proportional to the diameter of its lens. In other words, to see something small a long way off you need a very big telescope.

Apollo 11's Eagle lunar module measures about 4.3 metres across, and to see it from Earth, when we are at our closest to the moon, would require a telescope with an angular resolution of 670 billionths of a degree. If we take the wavelength of the reflected light from the moon as being 550 nanometres, the middle of the visible range, then to see the lunar module would require a telescope with a diameter of nearly 60 metres. The largest telescope now in existence, the Gran Telescopio Canarias on the Spanish island of La Palma, has a diameter of 10.4 metres.

Larger telescopes would be very expensive. The cost of building the European Southern Observatory's proposed Overwhelmingly Large Telescope, with a diameter of 60 to 100 metres, is estimated at €1.2 billion.

**Alby Reid**
Redhill, Surrey, UK

We can see distant galaxies but cannot see the much closer footprints left on the moon because galaxies and galaxy clusters represent a bigger target: they take up, or 'subtend', a much larger angle in the sky. Galaxies are also bright, making them stand out against the blackness of space. Footprints are simply impressions left on the lunar surface, offering no contrast at all. We would be reduced to looking for shadows cast by the tread.

Imagine two walkers about half a metre apart. In your mind's eye, draw lines from the focal point of your eye to the two figures. The angle between the two lines gets smaller as the figures walk away. The smallest angle at which the two figures can still be resolved is a measure of the resolving power of an optical instrument, in this case your eye. We use telescopes because they have a greater resolving power, so they can distinguish between objects that subtend a smaller angle.

With the naked eye – whose pupil has an aperture of about 2 millimetres – the two ramblers would blur into one object at a distance of about 2 kilometres, assuming perfect eyesight in which the ability to resolve two objects is limited only by diffraction. The best terrestrial optical telescope, the Gran Telescopio Canarias on La Palma, has an aperture of about 10 metres, giving it about 5,000 times the resolving power of the naked eye. A telescope of this power would be able to resolve our two ramblers even if they were 10,000 km away. However, the moon is 380,000 km away, and at this distance the telescope has no chance of separating the walkers, let alone their footprints.

To determine the resolving power of a telescope we use the Raleigh criterion. This tells us that the angle subtended by the smallest object an optical telescope can detect is roughly the wavelength of visible light divided by the aperture of the telescope. Multiply that by the distance to the object and we get the minimum size that can be resolved.

Taking visible light to have a wavelength of 555 nano-metres, the aperture of our terrestrial telescope to be 10 metres and the moon to be 380,000 km away, the smallest object that can be resolved on the moon would be about 20 metres across, assuming no atmospheric aberration. The ability to resolve footprints would require a telescope with an aperture of about 20 km.

If the Hubble Space Telescope were brought to within 40 km of the lunar surface it could achieve a resolution of 1 centimetre and make out footprints. Lunar-orbiting telescopes have come this close but their optics are not as good. Another option is to use an array of terrestrial telescopes to simulate a large effective aperture.

In any case I suspect that there are more interesting things to study, given that the sceptics would not be convinced anyway.

**Mike Follows**
Willenhall, West Midlands, UK

On NASA's website (bit.ly/PdSU) you can see the trail of footprints left on the moon by the Apollo 14 astronauts, photographed from the Lunar Reconnaissance Orbiter between 11 and 15 July 2009.

**Bill Watson**
Department of Mathematics and Computer Science
St John's University
Jamaica, New York, US

If the sceptics who doubt the moon landings don't believe the photos taken on the moon by the people who were standing there, why on earth (no pun intended) would they believe pictures beamed down from a telescope operated by the very organisation they suspect of lying to them?

**Stephen Gisselbrecht**
Boston, Massachusetts, US

# 6 Weird weather

## ? Rain imminent

*During the Monaco Grand Prix, I was watching the drivers battle with the damp conditions when the commentator said that rain was expected at the track in 6 minutes. How can forecasting be so accurate? If such technology is available, why isn't it offered to the public? Incidentally, the rain didn't arrive on this occasion, but was predicted with such confidence that presumably the forecasting must be accurate most of the time.*

**Derek Bolon**
Tetbury, Gloucestershire, UK

The reason we can achieve such accuracy in forecasting at each Grand Prix is because we have radar and weather stations on site provided by the the FIA, the governing body of world motor sport. In Monaco, these are operated by the French meteorological office, Météo France, which has an experienced forecaster on site to predict when inclement weather will arrive.

Each racing team is offered a subscription to the service, which is then fed to the team's timing stand at the trackside. This displays radar images, temperatures and pressures in near real time. There is also a minute-by-minute rain update provided via the trackside TV feed system, which also updates on-track events such as the blue warning flags waved at cars about to be lapped.

**Joseph Birkett**, Trackside IT Engineer
Red Bull Racing, Milton Keynes, Buckinghamshire, UK

*Many readers thought the answer was probably weather radar. The US, Australia, the Netherlands and Germany all appear to have publicly available services. Here's a typical example – Ed.*

I assume they use a weather radar. We have one sited 200 kilometres to the west of us, at West Takone, Tasmania, and by checking on the Australian Bureau of Meteorology website we can see a real-time picture of where the rain is falling and how heavy it is, or a loop showing how fast the rain is moving towards us. Great for getting the washing dry.

When our roof collapsed in the middle of winter and we were living under tarpaulins, it was wonderfully useful. The builders kept an eye on the weather radar, and when there looked to be at least 20 minutes clear they would whip off the tarps and work until I gave them a 5-minute warning to put them back again.

**Jan Horton**
West Launceston, Tasmania, Australia

# ❓ Chill factor

*Is it possible for it to be too cold to light a fire? In the same vein, is it possible for it to be so cold that a fire, with enough fuel to keep it going under normal circumstances, goes out?*

**Ian Tilly**
Ashford, Kent, UK

A fire is a rapid exothermic (heat-producing) chemical reaction which occurs between a fuel and an oxidant, usually oxygen. The rate of any chemical reaction increases with temperature, as the molecules move faster and collide more often. Most fuels and oxidants can coexist at room temperature without spontaneously igniting. Although they may slowly react together, the rate of reaction is so slow that the heat produced is dissipated before the mixture can heat up – think of a slowly rusting iron nail.

To start a fire, you need to heat the mixture to its ignition point. This is the temperature at which the rate of reaction is high enough to produce heat more quickly than it is lost to the surroundings. Heat energy starts to accumulate, driving up the rate of the reaction, producing even more heat, and so on. This runaway reaction is what we call a fire.

So the answer to whether it can ever be too cold to start and sustain a fire depends on the difference between the ignition point and the temperature of the surroundings. Heat is lost to the surroundings by a combination of radiation and convection. The colder the surroundings, the greater the rate of heat loss via these processes. Therefore, for low-grade fuels, such as wood, that do not produce much heat when they combust, a fire would not be able to sustain itself if the surroundings were cold enough. Instead, to keep it burning you would need a continual supply of heat from an external source.

However, there is a limit to how cold you can make the surroundings, culminating at absolute zero. So fuels with a high enough heat of reaction will always be able to sustain a fire, no matter how cold the surroundings. Conversely, even fuels that are normally difficult to ignite can be made to burn if the surroundings are hot enough.

**Simon Iveson**
Department of Chemical Engineering
University of Newcastle
New South Wales, Australia

# ❓ Surround sound

*I live a kilometre north of a busy motorway. When the wind is coming from the south the noise of the motorway is noticeably greater than when the wind is coming from the north. Assuming a wind speed of a mere 30 kilometres per hour, how can the wind direction affect the level of traffic noise I hear when the speed of sound is more than 1,235 kilometres per hour?*

**Jim Turton**
By email, no address supplied

Wind is the single most influential meteorological factor within approximately 150 metres of a noise source such as a highway.

The wind's effects are mostly confined to noise paths close to the ground. The reason for this is what is known as the wind shear phenomenon: the wind speed is lower in the vicinity of the ground because of friction.

This velocity gradient tends to bend sound waves downward when they are travelling in the same direction as the wind and upward when in the opposite direction. This process, called refraction, creates a noise reduction upwind from the source of the sound and a noise increase downwind from the source.

Over distances greater than 150 metres, vertical air temperature gradients are more important. This is because under certain stable atmospheric conditions, temperature increases with height either from the ground up, or from some altitude above the ground. Such an inversion occurs when a layer of warm air is trapped between layers of cold air. This inversion increases the speed of sound with increasing altitude, causing sound waves to be refracted back towards the ground. This would lead to an increase in ambient noise levels for far-away listeners.

**Victor Zeuzem**
San Mateo, California, US

The wind does not appreciably speed up the sound and, even if it did, this would not explain why the sound should be louder. What happens is that the sound is refracted, or 'bent', in rather the same way as a ray of light is refracted as it passes from air into water.

This happens because wind speed is not constant with height. At 100 metres altitude, say, the air is moving at 30 kilometres an hour. Closer to the ground, however, trees and buildings get in the way, so the wind speed is lower. At ground level, in between the blades of grass, the wind speed is close to zero.

Sound moving horizontally through air when there is a velocity gradient like this will be bent upwards if it is moving against the wind, and downwards if moving with the wind.

The best way to visualise this is to imagine a row of joggers with their arms hooked together running in a straight line on a beach. If the sand is uniformly firm, they all run together at the same speed and the line moves straight ahead. Now imagine that the sand is moist but firm at the end of the line of joggers nearest the water (providing fast-going conditions), and dry and soft at the other end away from the water (providing very slow running conditions). In this case the line will curve around because the fast runners have to stay hooked to the slow runners.

So, by the same reasoning, if sound travels 30 kilometres an hour faster at 100 metres altitude than it does at ground level, the sound wave front, which can be thought of as a planar disturbance, bends downwards.

**Hugh Hunt**
Department of Engineering
University of Cambridge, UK

## ❓ Hair-raising event

*Walking along the breakwater at Berwick-upon-Tweed in north-east England, my granddaughter and her mother noticed their hair was standing on end. It started to rain soon afterwards, but there was no thunder or lightning that day. What was happening?*

**Richard Turner**
Harrogate, North Yorkshire, UK

From one of my physics textbooks I recall a hair-raising picture of a woman standing on an exposed viewing platform at Sequoia National Park in California. She was in grave danger; lightning struck only minutes after she left, killing one person and injuring seven others (*Fundamentals of Physics*, 6th Edition, by David Halliday, Robert Resnick and Jearl Walker). It's likely that similar conditions were abroad on this day.

Most lightning clouds carry a negative charge at their base. Anything close to the cloud would feel the effect of electrostatic forces: electrons in a person's hair would be repelled downwards, leaving the ends of the hair positively charged. The positive hair tips then get attracted to the cloud – and repelled by each other – and stand on end. It's rather like rubbing a balloon on someone's hair to make the hair stand on end: the balloon becomes negatively charged and the hair is attracted to it.

Lightning victims often describe how they felt tingly and their hair stood on end before they got struck. Fortunately, air is a good electrical insulator and, in this instance, the charge in the clouds wasn't high enough to jump down to earth, so there was no lightning. However, this was probably a lucky escape for your family. If your hair stands on end outdoors or your skin is tingling, lighting may be imminent and it's best to run for suitable shelter.

**Iain Longstaff**
Linlithgow, West Lothian, UK

The phenomenon described above is known as luck – the two people were fortunate not to have been struck by lightning. Experienced hikers and climbers know that this hair-raising phenomenon can be a precursor to a lightning strike and are taught to flatten themselves or, if climbing, dive for lower ground.

There is a vertical voltage gradient in the atmosphere, typically in the order of 100 volts per metre on a clear, dry day. For an average adult male, then, there will be a 180 to 200-volt difference between the toes and the top of their head.

When electrically charged would-be storm clouds scud overhead, an induced ground charge follows the clouds, markedly increasing the voltage gradient. If the potential difference is sufficient to overcome the resistance of the air – around 3 million volts per metre – then lightning leaps across the gap. In practice, lightning strikes are possible at substantially lower voltage differences. The fact that the reader saw no lightning and heard no thunder merely suggests that, luckily, the voltage never rose high enough for a lightning strike.

**Larry Constantine**
Department of Mathematics and Engineering
University of Madeira
Funchal, Portugal

# ❓ Apple melt

*The snow at the base of our small apple trees melts before snow elsewhere has melted. We've seen this under other trees too. Why?*

**Robert Campbell**
Uetendorf, Switzerland

Snow, like everything else, including apple trees, emits and absorbs radiation. While ultraviolet and visible radiation are strongly reflected (not absorbed) by snow, it is however a strong absorber of infrared radiation. The battle between the absorption and emission of radiation determines whether there is net warming or cooling of the snow – or neither.

So why would snow under a tree melt faster? At night, snow in the open absorbs infrared radiation from the ground and from the sky – which can be below -30 °C when it is clear.

Snow underneath a tree absorbs radiation emitted by the ground and by the tree, which is likely to be significantly warmer than the sky, so it emits more infrared energy.

This difference is sufficient to explain why snow underneath a tree might melt faster than snow of the same depth that is out in the open, and also explains why frost often does not form around trees.

It is also possible that shelter provided by the tree when the snow was falling led to a thinner layer of snow there than in the rest of the immediate vicinity!

**Thomas Smith**
Environmental Monitoring and Modelling Group
Department of Geography
King's College London, UK

# ? Happy not sad

*Some years ago, between rain showers, I noticed an upside-down rainbow (u-shaped rather than n-shaped). The colours were also reversed. It appeared around 40 degrees above the horizon and was smaller than an upright rainbow. It persisted in a semicircle for about a minute before slowly fading from one side, the remaining arm lasting for another minute. Can anyone explain this?*

**R. Dufton**
Banbury, Oxfordshire, UK

Rainbows are indeed circular but intersect with the ground before their full circle can be achieved, giving an arc shape. As this rainbow was high in the sky (40 degrees above the horizon), we can assume it's possible other factors were involved, including reflection.

**Michael Edie**
Petersfield, Hampshire, UK

This is possible if the viewer has a reflecting surface, such as a sheet of very calm water, behind them. The reflection of the Sun from this surface can produce a rainbow in front of the viewer. Because the Sun is reflecting from water behind the viewer, if the viewer were to turn around and observe the Sun's reflection in the water it would appear as though it was beneath the surface of the water. The centre of the Sun would, therefore, appear to be below the horizon. A full rainbow circle could be produced under these circumstances appearing in front of the viewer as the Sun shines up into the sky rather than the usual sunlight which shines down. What was seen in this case was a part circle comprising the lower half of the circular rainbow.

**Malcolm Brooks**
The Met Office Press Office, London, UK

This radially challenged and disoriented rainbow sounds like a portion of a solar halo. This effect is caused by refraction of sunlight through a thin cloud of ice crystals such as that found in a cirro-stratus cloud veil. Various circles and arcs can be produced. This particular sighting would depend on the movement and extent of the veil and the rain clouds below it, relative to the Sun. Similar, but less colourful, haloes can be seen in moonlight.

Even smaller rainbows can be produced that are concentric with the Sun and Moon, by reflection and diffraction of light by water droplets in lower clouds. This effect is called a corona, which is not the same as the Sun's corona in an eclipse. The fact that lunar haloes and coronas appear to lack certain colours is, I suspect, linked to the human eye's reduced capability to discriminate colours in low light conditions.

**Steve Mason**
Eastleigh, Hampshire, UK

U-shaped rainbows are quite simply ones originating deep in the southern hemisphere. They only occasionally migrate north on exceptional weather systems. For one to persist as far north as Oxfordshire is probably a record.

**G. W. Storr**
Bournemouth, Dorset, UK

# ❓ Drip dry

*If I alight from the bus and it is raining, I tend to run for my
door, in the belief that I will arrive home less wet than if I walk.
However, I have heard that the same number of raindrops will
strike me whether I run or walk. Is this really the case?*

**Mark Haines**
London, UK

The volume of space swept by the body between bus and door
is identical. Therefore, assuming a constant rate of deluge,
the number of falling rain drops (below the crown of the
head) swept is the same. However, the number of raindrops
falling directly on to the top of your head is proportional to
the time spent exposed to the rain, so running reduces this
component.

But running will deposit all the swept component in a
shorter time. This will produce greater apparent wetting since
normal evaporative drying has less time to work. So for light
showers, with small swept and falling components, walking
is probably preferable. We make this complex decision
completely unconsciously, while also taking into account the
likelihood of the rain becoming harder or lighter, the distance
we have to travel, and our ability to run.

It would be interesting to confirm this theory by filming
pedestrians, recording the rate of rainfall, and relating the
latter to the point at which the former begins to run.

**Mike Stevenson**
Millom, Cumbria, UK

If the walker is in the Lake District, where horizontal rain is
common (and always opposes the direction of travel), then
it is recommended that they move as quickly as possible
because the volume of drops swept out through the rain is

now determined by the relative velocity of rain and walker multiplied by the time taken on the journey.

Indeed, if the rain is moving over the ground at speed $v_r$ (opposing travel) then the walker, moving at speed $v_p$, will be $1 + v_r/v_p$ wetter than in vertical rain by the time they reach shelter. By running to keep up with the rain (defined as $v_p = -v_r$) it is theoretically possible to stay dry.

**Martin Whittle**
Sheffield, South Yorkshire, UK

The following (from memory) is attributed to one D. Brown of York:

> When caught in the rain without mac,
> Walk as fast as the wind at your back,
> But when the wind's in your face
> The optimal pace
> Is as fast as your legs can make track.

**Matthew Wright**
University of Southampton, UK

# ❓ Getting the drift

*Why is it that snowstorms can last a long time and precipitate a deep layer of snow, but hailstorms are brief and do not result in a deep layer of fallen hail?*

**Mathias Brust and Robert Wilson**
University of Liverpool, UK

Most snow falls from extensive banks of stable nimbus clouds. The genesis of a snowflake requires fairly slow and continuous crystallisation with little turbulence. If the wind continues to carry clouds towards high ground, or if a warm air mass slides smoothly over a cold one, snow can persist for as long, and over as large a continuous area, as ordinary rainfall.

Hailstones are formed in conditions of violent convection inside tall, isolated, unstable cumulonimbus clouds, where there is very strong internal circulation. Most hailstones have a layered structure, showing that they have been cycled several times between freezing and melting levels. Eventually the cloud grows to the point where its shadow prevents the sun from heating the ground which normally supplies the rising thermals necessary to keep the stones airborne, and so they fall in a brief, localised burst.

Deep layers of hail do occur but, as the largest cumulonimbus clouds are formed on clear sunny days over hot ground, and as the same cloud may also precipitate warm rain, hailstones tend to melt quickly. As hail is macroscopically denser than fresh snow, the same mass of water will produce a much deeper layer of snow over a given area.

**Alan Calverd**
Bishop's Stortford, Hertfordshire, UK

Very severe hailstorms, such as those that cause crop damage in the summer in North America, can produce a deep layer of fallen hail. I have seen such a storm in Nebraska deposit about 10 centimetres of 5-millimetre diameter hailstones over a wide area in less than 20 minutes. At first glance, the appearance on the ground was of heavy snowfall. Of course, the storm moved on, the sun came out and the hailstones melted. In the UK our thunderstorms are not so severe and we only rarely and briefly see a coating of hail on the ground.

**Clive Saunders**
Manchester, UK

It is not necessarily true that hailstorms cannot produce a deep layer of fallen hail. Last summer, on a hot, sultry day, we had a thunderstorm that lasted for several hours. In the course of it, there was a fall of enormous hailstones (the largest I have ever seen) and they piled up in impressive drifts. Obviously, it was too warm for them to last for a long time on the ground, and they melted, producing a flash flood. Our garage was flooded to a depth of about 5 centimetres as a result.

**Diana Brown**
Etagnières, Switzerland

# 7 Troublesome transport

## ❓ Wheels of death

*I heard that the car is the deadliest weapon created by humans and that the number of lives it has claimed exceeds the death toll from atomic weapons, guns or bombing. Is this true? And what are the grisly figures involved?*

**Thomas Elling**
London, UK

First, we have to assume that this comparison sets automotive fatalities against all uses of weaponry, including acts of war. If that is the case, weapons win hands down.

This is for a number of reasons, the first being that, unlike weapons, the automobile was not designed with efficiency of death in mind: most road deaths are accidental. Furthermore, while spears, guns and explosives have been available for centuries, automobiles have only been around for about 120 years. They have only been in mass production for 100 years and accessible to most of the world's population for 60 years.

So what are the numbers? On the roads of the USA there has been an average of between 40,000 and 50,000 fatalities annually since 1970. If I were to extrapolate those numbers over 100 years (which would be to claim 50,000 died in years when there were barely 50,000 autos in the USA), then double the numbers again to try to include Europe, Russia, Japan and Australia, I would come up with slightly more than 10 million fatalities over the century.

In contrast, during the Second World War alone, combat deaths have been estimated at around 20 million. Civilian deaths by weaponry – including bombing and atomic bombs, but excluding the Holocaust, famine and other events – could probably total 20 million. I would argue that, unless millions of fatalities in remote lands have gone unreported, the allegation incriminating the car is unfair.

However, excluding warfare from the calculation would at least create a debate. Firearm deaths in the USA in 1999 totalled 28,874, of which more than 16,500 were suicides, 10,800 were murders and the rest were accidental or undetermined. According to the US National Safety Council, motor vehicle deaths in that year totalled 42,401.

**Alexander D. Mitchell**
Baltimore, Maryland, US

Although the statement concerning the relative lethality of motoring compared with warfare is a canard, like some myths it does have a kernel of truth. It originated during the 1980s in revisionist historical reassessments of the US involvement in the Vietnam War, when it was claimed that more young men were killed each year on American roads than died fighting in the jungles of south-east Asia.

In fact, during a decade of fighting, losses by the US armed forces totalled 47,378 – only slightly more than the average of 45,000 people killed *each year* in automobile accidents on American roads during the mid-sixties. Ironically, most of the 10,824 non-combat fatalities that US forces suffered in the conflict have been attributed to some kind of vehicular accident. Moreover, the highest casualty rate in both Vietnam and on the roads occurred in the same group: men in their late teens and early twenties. So from a revisionist perspective, going to war was almost 10 times as safe as driving a car.

Even if there were a basis for comparison, the Vietnam

casualty factor was quite specific to the US armed forces. For example, the total death toll inflicted on the indigenous population – civilian, military or insurgent – during the Vietnam War was between 12 and 13 per cent which, had the US population suffered proportional casualties, would have left 28 million Americans dead.

**Hadrian Jeffs**
Norwich, Norfolk, UK

The table fork is by far the deadliest weapon created by humans. Each year, this humble utensil abets the deaths of millions of people by conveying into their bodies all kinds of fatty foodstuffs known to cause heart attacks, cancers, strokes, diabetes and other diseases. According to the World Health Organization, approximately 17.5 million people died of cardiovascular disease alone in 2005, making up 30 per cent of all deaths globally.

As most of these harmful foods are of animal origin, and because the question doesn't specify human lives claimed, we might also add the number of animals killed to be eaten with forks to the yearly death toll. This amounts to about 56 billion, says the Humane Society of the United States.

**Jonathan Balcombe**
Physicians Committee for Responsible Medicine
Washington DC, US

According to several research studies, the US death rate due to medical misadventure is around 225,000 deaths per year, made up of 12,000 deaths due to unnecessary surgery, 7,000 from medication errors in hospitals, 20,000 caused by other errors in hospitals, 80,000 fatalities from infections in hospitals and 106,000 deaths due to the negative effects of drugs. So arguably the most lethal invention is in fact a doctor.

**Jeremy Ardley**
Perth, Western Australia

By not considering the less developed countries, some of your previous correspondents have neglected nearly 90 per cent of deaths caused by road traffic accidents. The World Health Organization estimates the total is 1.2 million deaths per year, the majority of these probably being pedestrians and cyclists.

It is also necessary to add the deaths from 'traffic-related air pollution', as another WHO report terms it. These have been estimated as between 0.8 times (in New Zealand) and three times (in Switzerland) as high as the accident-related death rate, so perhaps it would be correct to double the estimate of deaths due to motor vehicles to 2.4 million.

We should also factor in an estimate for deaths due to climate change caused by vehicle and petroleum industry carbon dioxide emissions. A definite value for this number of deaths will presumably remain unknown until it starts to reach tens of millions per year, because it is difficult to attribute individual droughts, storms or floods to climate change.

Even disregarding climate-change deaths, however, the current annual rate of deaths due to motor vehicles is at least double that caused by war.

**Philip Ward**
By email, no address supplied

# ❓ Sick as a horse

*On a long motorway journey while driving behind a horsebox, I wondered, do horses get travel sick? In fact, do we know whether any animals besides humans suffer from motion sickness?*

**Neil Bowley**
Newthorpe, Nottinghamshire, UK

Horses are unable to vomit, except in extreme circumstances, because of a tight muscle valve around the oesophagus. So it is difficult to know whether or not they feel sick. Other monogastric animals can vomit. Younger cats and dogs frequently vomit during their first car journeys but rapidly become accustomed to travel and no longer suffer sickness. In the UK a neurokinin-1 receptor antagonist has recently been licensed as a treatment for motion sickness in dogs as it reduces the urge to vomit.

**James Hunt**
Taunton, Somerset, UK

Motion sickness is common among animals, affecting domestic animals of all kinds. A carsick dog is not only pathetic, but messy. In his unforgettable book, *A Sailor's Life*, Jan De Hartog wrote: 'My worst memories of life at sea have to do with cattle. Two things no sailor will ever forget after such an experience are the pity and the smell... cattle get seasick, and the rolling of the ship terrifies the wits out of them. A seasick monkey or pup may be amusing and easy to deal with, but five hundred head of cattle in the throes of seasickness are a nightmare...' He also mentioned horses explicitly and even fish transported in unsuitable conditions may show signs of disorientation.

Motion sickness is ubiquitous because all vertebrates have organs of balance and they correlate balance with feedback from other senses to stay upright. When movement causes, say, visual information to conflict with balance, the brain of a sensitive individual interprets the disorientation as a symptom of poisoning and a typical reaction is to vomit to clear the gut.

**Jon Richfield**
Somerset West, South Africa

Both Robert Falcon Scott and Ernest Shackleton took ponies with them to Antarctica. On the way they experienced some appalling weather, and both noted how badly affected their animals were. They did, however, perk up when the storms abated. Similarly, Scott's dogs spent most of the storms curled up or howling, suggesting they too were suffering. Animals with a similar auditory system to ours would suffer from motion sickness, because it is caused by the confusion of auditory and visual inputs.

**Tim Brignall**
Bristol, UK

# ? Aerial glue

*I heard that a Formula 1 car travelling at 200 kilometres per hour would generate enough downforce (or suction) to allow it to stick to the ceiling. Is this correct? And if it is, how is the force generated?*

**Robert Webber**
Melbourne, Australia

*This is an interesting thought-experiment and gave us some great answers. Clearly the long-running debate over how aircraft wings achieve lift is still alive and kicking – Ed.*

The short answer is 'yes', and downforce is the way to do it. Downforce acts towards the road, whatever the road's orientation, and it increases roughly with the square of the vehicle's speed. Driven fast enough, the downforce exceeds the weight of the car, which could then run along the ceiling of a tunnel. Depending on the set-up, the downforce and the weight of an F1 car typically become equal when the car is running at 130 kilometres per hour.

Filling a wardrobe with clothes boosts its weight. This increases the friction between it and the floor, making it harder to slide. An F1 car designer wants to increase the friction between the tyres of a car and the track so that it can carry more speed into corners without sliding off. But the designer wants to achieve this without increasing its weight.

So downforce is the answer and it can be achieved in two ways. First, upside-down wings are angled to deflect air upwards, away from the track, resulting in a reactive force on the car in the opposite direction. Second, designers exploit the Bernoulli effect. Pass air through a narrowing gap and it speeds up. This is what happens beneath an F1 car because the space between the ground and chassis represents a constriction to the airflow. According to Bernoulli's principle, this leads to reduced pressure under the car. The ambient pressure above the car is higher than that beneath it, leading to a net force in the direction of the road.

Tunnels would need to be adapted to allow racing cars to run along the ceilings, which would make for some interesting overtaking manoeuvres on street circuits with underpasses like Monaco. It would also lead to spectacular crashes: any car that braked heavily while running along the ceiling would lose its downforce – or 'upforce' in that scenario – and fall, upside down, onto the track below.

**Mike Follows**
Willenhall, West Midlands, UK

In theory, yes, an F1 car could drive upside down – but only in theory. Racing cars generate a substantial part of their downforce by creating a low-pressure area under the car. This was most apparent in the late 1970s, when the cars started using skirts that went all the way to the ground in order to contain the low-pressure area. The Brabham F1 team even built a 'fan car' that sucked the air from under the car – ostensibly to cool the engine – creating a low-pressure area underneath. It was banned after its first and only victory, in part because it was impossible to follow closely because it spat stones and other debris out of the back into the faces of oncoming drivers.

Three factors are key. Is the ceiling strong enough to withstand the force of the car pushing against it? Is it flat enough to allow the car to run close enough to the ground to generate the required downforce? And finally, you would have to get the car up there in the first place. No doubt if someone built a tunnel long enough, a car could drive up using a curved ramp built into the walls.

Then there's the problem of grip. Because the car is being held up by airflow, this is the only thing pushing the tyres against the ceiling, so the driver would have very little control over the car. It would be difficult to generate braking without the wheels locking up and, if the driver tried to turn the car too quickly, it would skid. Then, if it were no longer facing forwards, it would lose its 'upforce' and fall. Because the car relies on the tyres for directional stability, and has no control surfaces like an aeroplane, the driver would be unable to 'land' the car back on the ceiling again.

**Marco van Beek**
London, UK

F1 cars are capable of generating up to three times their weight in downforce. An F1 car has wings, but these are mounted upside down, generating negative lift which pushes it against the ground.

Ground effect is also used; the underside of an F1 car is completely flat and the closer it is to the ground (its ride height), the smaller the gap becomes. Air travelling under the car is forced to travel faster than air above it, generating further negative lift. Ride height is regulated in Formula 1 because if the car bottoms out or hits a bump the downforce and traction are lost. Such an incident was claimed to have been a contributing factor in the crash that caused the death of former F1 world champion Ayrton Senna in 1994.

Nevertheless, the aerodynamic design and relatively low weight of an F1 car would, in theory, allow it to race upside down.

**Michael Macpherson**
Automotive engineering student
Glasgow, UK

# ? Speed freaks

*A single Formula 1 car passing by makes a noise of around 110 decibels. Last year I went to a Formula 1 Grand Prix and sat near the start line, where 20 cars left the grid at once. The noise was mind numbing, much louder than a single car, but not 20 times louder (or 2,200 decibels, an unachievable figure). Why wasn't it?*

**Walter Coppin**
Edinburgh, UK

There are two things here. First, the sound sources are not synchronised: the cars are making noise independently of one another, and so a burst of loud noise from one might coincide with a drop in noise from another. For these unsynchronised sound sources, the sound pressure (the air compressions that your ear detects) increases only according to the square root of the number of sources.

Second, the decibel (dB) scale is logarithmic: things don't just add together neatly. The increase in the decibel level is the logarithm of the ratio of the number of sources (in this example, 20/1) multiplied by 10. The 19 extra cars add only around 13 dB, so the noise from 20 cars will be just 123 dB or thereabouts.

The same is true for violins in an orchestra. The 16 violins in an orchestra produce only four times as much volume as a single violin: if one violin produces 70 dB, 16 produce 82 dB. Similarly, silencing half of the trumpets – which are obviously much louder than violins – only reduces their volume by a few decibels, which explains why you need so many more violins than trumpets in an orchestra.

This also has implications for traffic-noise control. If the noise emanating from a car engine is roughly the same level as the noise from its tyres then there's not much point in reducing engine noise by more than about 3 dB without also reducing the tyre noise.

**Hugh Hunt**
University of Cambridge, UK

Back in 1990, I measured sound levels at a Formula 1 Grand Prix at Silverstone in the UK for health and safety. I found that the sound level from a single car passing, measured in the pits, was indeed about 110 dB.

The sound level varied widely throughout the race. In the first lap, all the cars passed by my measuring point virtually at once, but at the end all the cars were well spread out around the track. The difference in maximum level between the first lap and later laps where the cars were spread out was variable but around 12 dB, close to what theory predicts.

Loudness to a human observer is another matter altogether and differs from recorded sound levels. The human ear works using ratios, so doubling the sound power always produces roughly the same increase in loudness, no matter where you start.

A rule of thumb that works pretty well for most people is that an increase of 10 dB corresponds to 'twice as loud', so 20 cars passing at once would be a bit more than twice as loud as one car to a listener such as your questioner – the 13 dB given by the first correspondent.

This just goes to prove that the decibel is a very confusing unit of measurement. With this in mind I've taken on the challenge of explaining decibels for people who don't know what a logarithm is, at bit.ly/decibels.

**Tony Woolf**
Acoustic consultant
London, UK

# ❓ Point the way

*When the Apollo and other similar space capsules were returning to Earth it was important for the larger end of their bell-shape to face downwards. This is because the protective shield that resisted the intense heat created on re-entry by atmospheric friction as the spacecraft slowed was positioned there. How were the capsules designed so that they would always keep the larger, protective face towards the Earth and not flip over so that the pointed end faced earthwards? It seems to me that this would be likely to happen as this orientation would minimise air resistance. Or is my grasp of space flight a bit flimsy?*

**Bill France**
Leicester, UK

It is a common misconception that spacecraft entering the atmosphere do so while going straight down, towards the Earth. This is perpetuated by just about every space movie ever made. The truth is that the spacecraft are going nearly horizontal as they enter the atmosphere, even when returning from the Moon. They remain within 5 degrees of horizontal until they have lost 75 per cent of their speed.

It is the location of the centre of mass that determines which end points into the wind. In this case, it is very close to the heat shield. The centre of mass of the Mercury space capsules was aligned with its central axis and these craft made a ballistic re-entry, meaning there was no lift.

With the Gemini and Apollo capsules, the centre of mass was offset from the central axis. This made the heat shield tilt slightly so that it was not perpendicular to the relative wind. This provided a small amount of lift, which made re-entry a little longer but reduced the peak acceleration from between 10 and 12$g$ to around 3 or 4$g$.

**Stephen Wood**
Orlando, Florida, US

The orientation of an unguided body moving through a fluid depends approximately on the relative positions of the centre of mass and the centre of pressure. The centre of mass is the point about which the weight of the object would balance. The centre of pressure is the point about which aerodynamic pressures balance and, broadly speaking, the body will orient itself so that its centre of mass is ahead of its centre of pressure.

A classic example is an arrow. If you throw an arrow sideways, it will rotate until the head is foremost. This is because the heavy arrowhead places the centre of mass towards the front, while the fletching (or flight vanes) places the centre of pressure towards the rear.

The Apollo capsule was designed with the heavy

equipment cradled in the deep, rounded bottom of the spacecraft, while the crew compartment – much of which is empty space – was towards the pointed top. This placed the centre of pressure behind the centre of mass, which stabilised the capsule as it fell through the atmosphere. The centre of buoyancy (which is related to the centre of pressure) was also above the centre of mass, thus keeping the capsule upright as it bobbed in the sea after landing.

You can encounter a dangerous example of this with a poorly designed model rocket. If the rocket's fins are too small, or the mass of the engine and fuel too far to the rear, the centre of pressure will actually be ahead of the centre of mass. This will make the rocket highly unstable at launch, often spinning like a top as soon as it rises off its launch pad and tower.

However, as the fuel burns, the rear of the rocket will get lighter, moving the centre of mass steadily forward. If this moves the centre of mass ahead of the centre of pressure (where it should have been in the first place) the rocket will suddenly stabilise and start moving in a straight line, although in a random – and perhaps extremely hazardous – direction.

**Dan Griscom**
Melrose, Massachusetts, US

## ? The right path

*A pathway in my neighbourhood splits at an incline into both steps and a slope. Which option requires more calories to walk up or down?*

**Yonatan Silver**
Jerusalem, Israel

The answer is complicated. Stair-climbing is more efficient than walking up a ramp, so it costs less energy. What's more, ramps of the same steepness as an average flight of stairs are impossible to climb.

Very shallow slopes take a long distance (and walking time) to reach the same height as steep steps, so more energy is used overall. However, when climbing the steps, energy is used at a greater rate, which makes steps feel like harder work. In each case the gradient is important – steps with smaller risers and ramps with a shallower gradients feel less tiring because the rate or intensity of work is less.

**Michael Rennie**
Professor of Clinical Physiology
Derby City General Hospital
Derby, UK

# 8 Best of the rest

## ? Net rage

*When a thread or topic is started on a user-generated forum on the internet, it isn't long before one of the contributors makes a seemingly unprovoked attack on a total stranger. What is it about non-face-to-face contact of this kind that makes this more common than it would otherwise be?*

**John Anderson**
Warsaw, Poland

In 1987, psychologists Mary Culnan and Lynne Markus refined the 'reduced cues theory' to explain potentially abusive behaviour online.

They suggested that computer-mediated communication is inferior to face-to-face contact because social cues such as body language, tone, volume and intensity of speech are lacking. An online conversation, therefore, except when a webcam or microphone is used, takes place in what is termed a 'social vacuum'. The reduced cues that are available to each correspondent can lead to a lack of individual identity (deindividuation), which in turn undermines any social and normative influences.

Overall the lack of these strong influences can lead to forms of uninhibited and atypical behaviour. Behind a computer screen you are usually fairly safe from physical retaliation. This creates a sense of safety and a disguise for participants which is further reinforced by the control individuals can exert over their online identity.

On user-generated forums, for example, you can choose what profile information about yourself is displayed, fabricate that information, and in most cases choose not to disclose it to fellow participants at all. Similarly, in virtual worlds you can take on a name and an avatar which is entirely unlike the real you.

As to the motive behind an unprovoked attack, human beings are undeniably complex creatures: the reasons could range from simply having a bad day at work to wanting the excitement of causing trouble.

**Thomas Venus**
Nuneaton, Warwickshire, UK

Social interaction depends on innate and acquired attitudes, including the urge to be imposing, formidable or dominant. Contrary factors, such as fear, upbringing, affection or social pressures, tend to dampen down extremes of behaviour and prevent loss of control.

A healthy balance of all these structures one's behaviour in a socially desirable manner. Remove this feedback, and misfits, habitual victims of bullying or products of unhappy backgrounds revel in the freedom to indulge in bullying or sadism that has driven sensitive victims to suicide.

More sensible recipients of this kind of correspondence simply wipe off such nuisances in their 'kill' lists or otherwise 'kill them with silence', as the Japanese wisely put it.

However, people who indulge in abuse and bullying are widespread on internet forums, where they cannot be touched.

Other expressions of perceived immunity include football hooliganism in large crowds, and car drivers who feel safe insulting or threatening others. George Orwell characterised such impulses as 'the irresponsible violence of the powerless'.

Similar behaviour is common among animals, most familiarly lapdogs in vehicles, or safe behind high fences. They pose and threaten like monsters, but then panic abjectly if their protection fails and someone calls their bluff.

**Jon Richfield**
Somerset West, South Africa

What a stupid question, you total and utter... LOL!

**Fake Name**
By email, no address supplied

## ❓ Touchy feely

*In Olympic swimming events, the winner is the first person to touch a pressure-sensitive wall pad at the end of the pool. How does this pad know that a person has touched it rather than just registering the pressure of splashing water? If a swimmer just brushed it, would it fail to register their finish? I know that in the men's 100-metre butterfly event in the 2010 Olympics in Beijing, the equipment was called into question when Michael Phelps of the US won his seventh gold medal of the games. How did officials know it had operated successfully? And finally, it's easy to judge the victor in a race taking place out of water – such as running – because a sensor beam can scan the finish line. But in the pool how can they ensure that all the wall pads are exactly in line at both ends of the pool? Are they aligned before water is added to the pool and, if so, how?*

**Kelly Clitheroe**
Grimsby, Lincolnshire, UK

At the end of each lane there is a touch pad 90 centimetres high, 240 cm wide and 1 cm thick. Touching the pad stops the clock. Omega, the manufacturer of the touch pads used during the 2008 Olympics in Beijing, claimed that the pads react to the slightest touch from a swimmer's hand, but not to the splashing of water.

However, after the argument around Michael Phelps's victory over Milorad Cavic in the 100 metres butterfly final at the games, later verified by digital images, it was revealed that a pressure of approximately 3 kilograms per square centimetre must be applied to the pad to activate it and stop the clock. Therefore, it can be said that the victor is the person who touches the pad with enough pressure and not necessarily the one who touches the pad first.

**Joanna Jastrzebska**
North Shields, Tyne and Wear, UK

The pressure pad's tolerances are supposed to require a swimmer's touch before it will trigger a response. A pulse of water would have to come from a high-power nozzle to apply enough pressure to trigger the pad.

A swimmer approaching the end of a race cannot push a narrow enough or strong enough stream of water to trigger the pad. However, brushing the pad lightly may also not trigger it and so these days timing officials check overhead, high-speed cameras – like those used in track races – if the pad is just brushed or they are uncertain of the winner for any reason.

**Adrian Skinner**
Bournemouth, Dorset, UK

The pads are screw-fixed to the poolside along their top edge and in close contact with the poolside behind. The swimmer's positive and forcible pressure on the pad must close any gap between the pad and the poolside, or it may not register.

You have to hit the pad quite firmly to register, either at the turn or at the finish. Just occasionally the pad does indeed fail to register, either through poor swimmer contact or pad malfunction. This is why there are back-up timekeepers on each lane – both human and electronic – in order to verify a result. If a world or championship record is at stake, there must be at least three timekeepers present, and one of them has to be electronic.

The final published time may have to be scrutinised by the referee if there has been a mechanical problem; sometimes a compromise or average time may be recorded at the referee's discretion. Sometimes the record has to be disallowed if the electronic timing device is in any way compromised.

If a pad should malfunction during the course of a race, it is removed and exchanged between events, which takes about 5 minutes. The new pad is tested by punching it manually while timekeepers in a control room monitor the effect.

In major events, reaction times off the starting blocks are also electronically measured by sensors and displayed instantly on the scoreboard (this identifies false starts). The changeover time is also registered in relays, to show if the outgoing swimmer left the blocks before the incoming swimmer hit the pad. Relay swimmers still in the water while the race continues must take care not to touch any of the pads by mistake when they exit the pool, to avoid confusing the timing systems.

**Phil Sears**
Amateur Swimming Association club coach
Dorking Swimming Club
Westcott, Surrey, UK

# ? Jump start

*A sprint athlete is deemed to have false-started if they react within 0.1 seconds of the starting gun. This seems like a rather arbitrary round figure. What studies have been done to test human reaction times, and is the fastest a person can react to the sound of a gun really exactly 0.1 seconds?*

**Cathy Jameson**
Barrow, Cumbria, UK

The earliest scientific research into human reaction times was undertaken in 1865, by the Dutch physiologist Franciscus Cornelis Donders, best known in his lifetime as an ophthalmologist but who was responsible for pioneering studies in what became known as mental chronometry.

Donders measured response times by applying electric shocks to the right and left feet of his subjects. They responded by pressing, as quickly as they could, an electric telegraph key to indicate which foot had been shocked. In some tests the subjects were warned beforehand which foot was to be tested, in others no prior notice was given. By measuring the difference in reaction times between the two types of test, which he found to be 0.066 seconds, Donders made the first tentative calculation of the speed of a human's mental responses.

In the example raised by your questioner, where a starting gun is used, the figure established by the International Association of Athletics Federations, 0.1 seconds, is in line with the response time measured by Donders, rounded up to one decimal place. So, yes, that is almost certainly the fastest time an athlete, even after repeated practice, could respond to the starter's gun. If they react any faster, the clear inference must be that they were launching themselves off the blocks before the gun was fired.

**Hadrian Jeffs**
Norwich, Norfolk, UK

Many experiments have been carried out on reaction times. A number of these were performed by psychologists attempting to gauge correlations between reaction times and intelligence. Although a small positive correlation was found in this area by Ian Deary, Geoff Der and Graeme Ford of the Universities of Edinburgh and Glasgow, UK, in 2001, the issue is still contentious.

There are three main approaches to measuring reaction time. In the simplest of these, only one stimulus is presented to the subject, who is restricted to only one response. In the second type – called recognition reaction time experiments – several different kinds of stimuli are presented to the subject, who is required to respond to one kind and ignore the others; again there is only one correct response. Finally, in choice reaction time experiments, subjects are required to give different responses to different stimuli: for example, to press one button if a red light appears and another button if a green light appears.

Whichever approach is used, subjects normally perform several trials. The response times are then averaged to compensate for the variability from trial to trial, and so give a more reliable measure.

In the specific case of a starting gun, response times to auditory stimuli have been studied for a long time, and the generally accepted reaction time is approximately 0.16 seconds. However, there is a great amount of variation between individuals, as well as differences in single individuals over time.

**Ian Smith**
London, UK

# ❓ Ocker stopper

*In Australia, the loud, shrill call of 'Cooee!' is often used over long distances in rural and mountainous areas to draw attention to oneself. What is it about this word that makes it more audible over distance and is there a word more suited to drawing attention to myself should I be lost in the outback?*

**Tom Langford**
Brisbane, Queensland, Australia

*Plenty of speculation on this one so we're keeping the file open for a while. Thanks to those who point out that 'Gooweet' was devised by Australian Aborigines for gaining long-distance attention, while a few hundred years back 'hoooooha' was popular in Sri Lanka – Ed.*

A person's hearing is most sensitive between 1 kilohertz and 4 kilohertz, peaking at about 3 kilohertz. When someone hollers 'cooee', the highest part of the call is generally between 1 and 4 kilohertz. So it's not that the call itself carries better, it's just that we humans hear better at that pitch.

**Nate Dog**
By email, no address supplied

The word 'cooee' is more audible because it contains two open, long vowels. When pronounced, 'coo' opens up your throat allowing it to resonate. When the subsequent 'ee' is pronounced your throat closes slightly and the sound resonates loudly in your mouth and nasal cavity.

In contrast, when you say 'help', for example, this contains a very short vowel which only sounds in a narrowed throat.

So these 'big' vowels allow you to send your voice out louder and larger. As for words that will draw attention to yourself... any swear word with 'big' or 'long' vowels will do.

**Natalie Arad**
By email, no address supplied

The 'c', or any other explosive consonant, carries poorly and only helps to launch the subsequent vowel sounds. What you need to remember is that there are two very different vowel sounds which are at opposite extremes – one has almost no overtones, while the other is rich in them. Also, each vowel sound is sustained long enough not to be obscured by so-called multipath distortion, when different sounds reach the listener at the same time and thus confuse reception.

Similar considerations are apparent in the International Radiotelephony Spelling Alphabet (under which letters are known as 'Alpha', 'Bravo', 'Charlie', and so on for the purpose of clarity), and related procedural words such as 'Roger' and 'Mayday', which are chosen to be distinct even if you can only discern the vowel sounds. In fact, evidence shows that people use similar sounds when choosing a name for a dog; two distinct vowel sounds are best when calling it from a distance.

**Ron Davis**
By email, no address supplied

Australians may use this call to signal from afar, but Austrians yodel. In fact, 'cooee' resembles 'yodel(odel)ayitee' in that it contains broken, not continuous, sounds and also ends in 'ee'.

Comparative trials seem to bear out the use of these sounds, even taking into consideration climatic factors such as mist, rain and snow, and terrain such as bare rock against vegetation.

**P. G. Urben**
Kenilworth, Warwickshire, UK

# ❓ You cannot be serious

*How accurate can the automated tennis line-judging system called Hawk-Eye be? Surely for the level of accuracy it seems to offer, it would need far more cameras than appear to be present at major tennis tournaments. Yet everybody happily accepts its rulings. How does it work?*

**Franco Pallatini**
New York, US

Briton Paul Hawkins created and named Hawk-Eye, a system which combined the expertise he gained for his PhD in artificial intelligence with his passion for sport, particularly cricket.

In cricket, a batsman can be given out 'leg before wicket'. This ruling is applied when the umpire believes the ball would have gone on to hit the stumps had the batsman's leg not been in the way. In this situation Hawk-Eye can be used to predict the ball's trajectory and is arguably more reliable than an umpire.

Despite being invented with cricket in mind, it was tennis that was receptive to the technology much earlier, perhaps thanks to TV replays showing that umpiring mistakes contributed to the defeat of Serena Williams by Jennifer Capriati in the 2004 US Open quarter-finals. Hawk-Eye provides an instant replay of crucial shots and has also proved an excellent tool for analysing the strategy and performance of players.

For tennis, it relies upon a maximum of six cameras to provide data for sophisticated triangulation. The position of the ball is tracked via a succession of stills from each camera. Within a virtual recreation of the tennis court, a ray can be drawn from each camera through the centre of the ball. The intersection of these rays provides the position of the ball in three dimensions and, with the passage of each frame, its velocity. This can be used to calculate the contact area of the ball with the court, taking into account the distortion of the ball after it is hit. Hawk-Eye also captures any skidding of the ball on the court, which can deceive the eye into believing a ball is out.

**Mike Follows**
Willenhall, West Midlands, UK

## ❓ Call me for dinner

*I placed my mobile phone in my microwave oven, closed the door and then called it from a landline. I expected the oven to shield it from the incoming microwaves, but to my surprise the phone rang. Does this mean the oven is tuned or that it is leaking?*

**Tam Anderson**
Kirkcaldy, Fife, UK

Mobile phones and microwave ovens are designed to operate efficiently within a narrow band of radio frequencies. The microwave oven is tuned to 2,450 megahertz, which is 650 MHz higher than the highest band a European dual-band mobile phone can use and 550 MHz higher than an American phone. You might say that the oven and the phone are not on the same channel. The oven is designed to keep in all the energy it produces, in order to cook food. It does this well, thanks in part to regulations limiting leakage of the 2,450 MHz microwave energy that it uses. But at other frequencies – say the 900 MHz, 1,800 MHz or 1,900 MHz used by mobile phones, but for which the oven shielding is not designed – it might well leak energy in or out, which would permit a mobile phone to work from inside the oven.

**Michael Brady**
Asker, Norway

A friend of mine placed a mobile phone in the microwave and turned the oven on. This is not advisable. Because I am an electronic engineer, my friend then asked me if the phone could be repaired. It could not.

**By email, no name or address supplied**

I have made some interesting observations about microwave ovens.

In Alberta it is legal to drive with a microwave detector in the car. This, the manufacturers of the detectors tell us, is for our own safety, because the instruments alert us to the shower of microwave radiation which is emitted by emergency vehicles speeding up the freeway in order to save property and lives.

Perhaps of more interest to many owners of these delicate scientific instruments is that they can detect the radiation emitted by police vehicles intent on reducing the bank balances of speeding citizens. Out of concern for my safety, I carry one of these machines in my 260 horsepower automobile, and guess what? Every time I pass a supermarket which uses microwave ovens, the microwave detector goes off – about 200 metres away, in fact.

**Jamil Azad**
Calgary, Alberta, Canada

*The answer lies in the sensitivity and different tuning of your microwave detector and the shielding of the ovens. The ovens will leak some radiation across the spectrum – which you can pick up – but will not normally leak enough high-energy microwaves to make the ovens dangerous – Ed.*

# ⁉ Water substitutes

*Is water the ideal liquid in which to swim? Assuming there are no ill effects to your health, would a different liquid that was either denser and more viscous, or had some other property, enable one to swim faster or with less effort?*

**Robert Laing**
London, UK

The speed of a body in fluid is limited by the sum of three factors. The viscous drag is the friction of the fluid against the wetted surface. The form or pressure drag is the force created by the pressure difference between the front and the rear of the body. Finally, there is the wave-making drag, which is the energy wasted in making waves on the surface of the water.

I would suggest two strategies to achieve a higher speed for a given power. Swim totally submerged in a liquid with a lower viscosity and lower density than water, such as acetone, methanol or ether. The lower viscosity would cause less friction and reduce pressure drag, which is proportional to the density of the fluid, cross-sectional area of the swimmer and square of their velocity. Swimming below the surface would totally eliminate the wave-making drag. This is what submarines do to achieve high speeds.

The alternative is to swim on a liquid that has a much higher density than water but low viscosity. Mercury, which has a density 13.6 times that of water, would be ideal.

Archimedes' principle would ensure that just a small part of the body would be submerged, so the wetted surface and the viscous drag would be very small. The pressure drag would be about the same because while the cross-sectional area of the immersed part of the body would be reduced, the density of mercury is greater. The wave-making drag would remain. And of course you would have to invent a completely new style of swimming, probably something like paddling.

**Radko Istenic**
Ljubljana, Slovenia

Mercury has to be the best liquid in which swimmers can enhance their performance. A swimmer weighing 90 kilograms whose back has a surface area of 3,000 square centimetres could do a modified backstroke with their torso displacing less than one inch of the mercury.

The swimmer could keep all of their limbs out of the liquid and use vigorous heel kicks into the mercury as an effective means of propelling themselves forward.

Mercury does not wet skin and the sharp shape of its meniscus would further reduce the drag. With less than an inch of immersion, the swimmer's body would virtually 'hydrargyro-plane' across the smooth surface.

The swimmer could use hand strokes for further power and steering, but the technique would require experimentation as limb immersion could slow things down.

In a ceramic-fibre bodysuit a swimmer could do even better in a pool of molten gold, platinum or uranium, displacing barely half an inch of liquid – before frying when the thermal insulation of the suit failed.

**David Emanuel**
Tulsa, Arizona, US

# ❓ Freeze frames

*I spent time in the Scottish hills last winter and on a couple of occasions I had cause to clean my glasses in a stream that originated from melting snow, effectively at 0 °C. The water cooled the glass and its metal frame to such an extent that both lenses fell out. But how could this happen when, if I remember my school physics correctly, metal should contract more than optical glass because of a higher coefficient of expansion? Obviously this has never happened when I've been walking around under normal conditions.*

**Andy Douse**
Drumnadrochit, Invernesshire, UK

*We received some very entertaining answers to this question, but we haven't really nailed it yet. Several people called for more experimentation or wanted to know the coefficients of expansion for optical plastics so that they could be compared with those for metals. Do you know the answer? It's your big chance to appear in the next book – Ed.*

Your correspondent does not say whether he had put the glasses back on when the lenses fell out. If he had, the warmth of his body probably heated the frames more rapidly than the glass. If it happened while he was washing the glasses in the stream, then the rapidly falling temperature might have shrunk the metal frames and squeezed the lenses out.

**Doug Grigg**
Cannonvale, Queensland, Australia

Unless the questioner's spectacles were very old, his lenses would be made of plastic not glass. This has a coefficient of expansion many times that of steel.

**Alan Hickman**
Sleaford, Lincolnshire, UK

Glass lenses could not fall out of a glasses frame as a result of thermal contraction, but most lenses today are polymers. Steel has a linear thermal expansion coefficient of about $2 \times 10^{-5}$ per °C, while that for polycarbonate is about $7 \times 10^{-5}$ per °C. On cooling, the lens will contract inside the frame, but the total difference in contraction for 20 °C cooling of a lens 50 millimetres across would only be 0.05 millimetres. Assuming the lenses are properly located in the frame, this should not be sufficient to loosen the lens. I tested this by putting my reading glasses in the freezer – a 40 °C cooling – with no noticeable effect. Some more experiments may reveal why the lenses fell out.

**Philip Ward**
Sheffield, South Yorkshire, UK

# ❓ Splodge

*I was having a discussion with my mates about what would happen if you filled a swimming pool with jelly and jumped in. Some of the group believed you would sit happily on the surface. Others, myself included, reckoned you would sink, and risk drowning as the jelly collapsed around you. We wouldn't want anybody to be harmed, so we don't recommend experimenting to find out. But is there a theoretical answer to the question?*

**Ross**
Bristol, UK

The active ingredient in jelly dessert (or jello) is gelatin, a protein-based gelling product made from collagen.

Gelatin comes in different grades, or Bloom numbers, as measured by the force required to push a plunger into a solution of the stuff to a predetermined depth: the more rigid the sample, the higher the Bloom number. Jelly babies – a popular British sweet shaped like a miniature baby – have a high Bloom number, so there is little danger of drowning in a pool of the mixture used to make them.

The density of jelly is typically 10 per cent higher than water, so a swimmer would float higher in a pool full of jelly than in a pool full of water. Jelly is also more viscous than water, meaning that someone diving into jelly might have difficulty surfacing. However, two researchers from the University of Minnesota, Minneapolis, won the 2005 Ig Nobel prize for chemistry for showing that people could swim just as quickly in water spiked with guar gum, an edible thickening agent, as in ordinary water. The spiked liquid has double the viscosity of water, yet the increased drag is cancelled out by the increase in thrust that swimmers can generate in it.

While we're on the subject of desserts, custard is interesting, as it becomes much more viscous under pressure. It is possible to walk across a pool full of the stuff, as demonstrated on the UK TV series *Brainiac* (for a clip of the feat, see bit.ly/3dkkg).

**Mike Follows**
Willenhall, West Midlands, UK

Jelly is an interesting substance because it behaves as both a solid and a liquid. We did some simple experiments to pin down its behaviour.

We found we could stack five 20-cent coins on top of some jelly before they punctured the surface, meaning the jelly was able to support a pressure of 700 newtons per square metre. This is only around one-tenth of the pressure a person would exert if they tried to sit on a pool of jelly, so presumably such an attempt would fail.

Once the surface broke up, we found the jelly behaved as a very viscous fluid, flowing around a small capsule we pushed into it. From the force required to move the capsule, we estimated the jelly's viscosity to be around 50 pascal seconds (50 times that of castor oil).

This is so large that a person jumping into a pool of jelly would be slowed dramatically in the first fraction of a second – doing a belly-flop onto jelly would hurt. They would then sink gradually until their buoyancy became neutral, although they would float higher than in water as jelly is a little denser.

**Finn Lattimore and Ruth Mills**
Canberra, ACT, Australia

## ❓ Coins of the realm

*Photographs of the clock mechanism in Big Ben, the face of the Houses of Parliament in London, show coins on a tray attached to the pendulum being used to tune the oscillation period. Since we all know that mass doesn't affect the period of a pendulum, why is this done?*

**Tony Richardson**
Ironbridge, Shropshire, UK

Your questioner is correct in saying that the mass of a pendulum does not affect its period. However, the length of the pendulum does, specifically the distance between the pivot and the pendulum's centre of mass. Adding coins to the pendulum moves the centre of mass slightly, changing the period.

**Gareth Shippen**
Bromsgrove, Worcestershire, UK

A pendulum swinging under the force of gravity is an example of simple harmonic motion. It is easy to work out that the period is proportional to the square root of its length but independent of the pendulum's mass.

A clock pendulum is designed to have a period appropriate to the train of gears linking (in most cases) the minute hand with the 'impulsing escapement', which converts the oscillations into rotational motion. This period is associated with the length between the pivot and the centre of oscillation, where all the pendulum's mass is concentrated. Addition of mass above the centre of oscillation will raise this centre, shortening the period. The converse also applies.

In the 1990s I had the pleasure of working for the clock-making firm Thwaites & Reed, then responsible for Big Ben's maintenance. The clock's actual name is the Great Westminster Clock; strictly speaking, 'Big Ben' refers to the hour bell. The clock's pendulum has a 4-second period and its centre of oscillation can be calculated to be about 3.97 metres from the pivot. The addition of a penny to the tray on the pendulum raises the centre of oscillation by 0.0368 mm – enough to cause a gain of 0.4 seconds per day. Removing a penny has the opposite effect.

The pendulum is complex, comprising a steel suspension spring, a steel/zinc/steel temperature-compensating structure, and a cast-iron 'bob' attached to the arm, with an adjustable rating nut at the bottom to regulate the swing. The bob swings 3 degrees (about 190 mm) to either side of the vertical. The pendulum's length, including the spring, is about 4.5 metres and the whole construction weighs 322 kg. It is obvious that it is impractical, even dangerous, to stop the pendulum, make an adjustment to the rating nut and restart it. The disturbance to the pendulum, and to the clock's time-keeping, would be intolerable.

Edward John Dent and his adopted son Frederick, who built the clock in the 1850s, built their large clocks to run fractionally slow so they could easily be made to tick at the correct rate by adding weights to a top tray. Pennies were used because of their predictable response.

For small swings, a pendulum's period is independent of the size of its swing, but in Big Ben's case the pendulum has a large amplitude so this does not hold. Errors of up to 13 seconds a day could occur if the amplitude changed significantly for whatever reason.

That said, I analysed Big Ben's performance over several years and found it is well able to maintain accuracy of less than 1 second gain or loss per day, more than 95 per cent of the time. Accuracy is maintained by noting and acting upon the difference, if any, between the first strike of the hour bell and the time from the 'speaking clock' service provided over the phone. Records show the clock does not appear sensitive to changes in temperature or atmospheric pressure, responding only to the addition or removal of pennies as the clock's attendants correct random drifts in timekeeping.

So what causes these drifts? Well, there is evidence that the zinc core supporting the bob has been deformed under its load so the position of the bob is lower now than when the clock was built. This has been corrected over the years by increasing the mass on the tray. This stretching means that the pendulum's temperature compensation is not as good as it once was, but there have been no obvious effects. Any effects arising from changes in atmospheric pressure are likewise too small to observe.

The *British Horological Journal* (vol. 151, p. 437) describes a pendulum clock designed by horologist Philip Woodward and owned by a Californian, David Walter, that is sensitive to seismic tremors imperceptible to humans, and probably to traffic on a nearby California freeway as well. In a similar spirit, I would speculate that Big Ben can feel the twice-daily ebb and flow of the Thames, as well as vibrations from the tube trains passing underneath, not to mention MPs moving in and out of the chamber of the House of Commons just to the south.

**John Warner**
Burgess Hill, West Sussex, UK

# Index

**A**

adipose tissue, as insulation 39
adrenalin surges 40
air spaces
    and floating/sinking citrus
        slices 23
    in peppers 25
albedo (pith), and floating/
    sinking citrus slices 24
alcohol
    development of tolerance to
        beer 52–4
    downing a pint in one go
        15–16
    level of consumption 76–7
    and mood 70–71
    'shaken not stirred' vodka
        martinis 6–9
alcohol dehydrogenase (ADH) 52
aldehyde oxidation, in vodka
    martini 6, 7
alkalaemia 163
alloparental care 113
allyl isothiocyanate, and hotness
    of mustard 12
alternating current (AC) 97, 99
analogue technology 80–81
animals killed for food 191
antigravity effect 171
anxiety, physical reactions to 40
Apollo spacecraft
    Apollo 11 172
    Apollo 14 174
    orientation on returning to
        Earth 199–201
appletini garnish, buoyancy of
    10–11

Archimedes' principle 215
ash levels, in dog food 141–2
astigmatism 133, 136
athletics
    athletes competing at high
        altitudes 162–4
    athletes' variations in form
        58–60
    reaction times to starting guns
        208–9
atmospheric pressure 162–3
automatic activity, and thinking
    about actions 37–8
avatars 204

**B**

background noise 84–5
bacteria, wet items 87
balance system 75, 193–4
ballpoint pens, vented to prevent
    leakage 88–9
Basidiomycetes fungi, medical
    benefits of 14
Basilosaurus 117, 118
bath temperature 64
bats
    echolocation 150, 152
    hanging upside-down 150,
        151
    heads kept stable in flight
        151
    sacculus 151
    semicircular canals 151
    vestibular system 151, 152
batteries
    lithium 103
    nickel-cadmium 102

nickel-metal hydride (NiMH)
102, 103
beer
development of tolerance 52–4
downing a pint in one go
15–16
beetles, landing on their backs
130–33
bell curve 59–60
Bernoulli effect 195
Big Ben 220–23
biosonar 152
biosphere 145–6, 159
birds
birdsong 118–19
black-chinned hummingbirds
139
and colour coding of milk
bottle tops 109–10
recognition of own young
112–13
ruby-throated hummingbirds
138–40
standing on one leg 120
black-chinned hummingbirds 139
Blackpool Tower, and curvature
of the Earth 155–7
blinking 73, 74
blood alcohol concentration
(BAC) 70–71
blood pressure
increase in 40
reduction in 40
Bloom numbers 218–19
blowholes 116–18
blue tits, and colour coding of
milk bottle tops 110
brain
brain volume and transsexual
people 49
correction of inverted images
41–3
interpreting the acoustic
environment 84, 85
problem-solving 57
storage capacity 47–8
unlike a standard computer 47

'broken heart' 39
BSE 142
bubbles
and buoyancy of appletini
garnish 10
and cloudiness in shaken
martinis 6, 7, 8
in still mineral water 96–7
and water density 21, 22
buoyancy
of appletini garnish 10–11
jumping into a pool of jelly
220
of spacecraft 201
burials 55
butterflies, height of flight 124–6

C
camouflage 128–30
cannibalism, and monodiet 4, 5
capsaicin, and hotness of chilli
peppers 11–12
capsicums *see* peppers
*Carabus nemoralis* 132
carbon dioxide
in a biosphere 145, 146
emissions 192
at high altitudes 163
in peppers 25–6
and water density 21
cardiovascular disease 191
carnivores 129
cars
car fatalities compared to
those from weaponry
189–90, 192
deaths due to climate change
caused by carbon dioxide
emissions 192
Formula 1 car downforce
194–6
Formula 1 sound levels 197–9
ground effect 197
insulting/threatening drivers
204
road traffic accidents 192
traffic-related air pollution 192

carvone 51
catalase 52
cataracts 136
cats
    drinking method 123
    intolerance of cold water 122
    and motion sickness 193
cellulose 82
cephalopods 144
cetaceans 116, 117
chelating agents 89
chemosynthesis 160, 161
chilli peppers, duration of
    hotness 11–12
chloroplasts, in peppers 24, 25
choice reaction time experiments
    209
chromatophores 144
chromoplasts, in peppers 25
circumnutation 121
cis-trans isomerism 51–2
citrus slices, floating/sinking
    23–4
classical music, noise levels 69
click beetles 131
climate change 192
clitoral sensitivity 68
cloaca 29
clock pendulums 220–23
clouds, albedo of 165, 166
coefficient of expansion 217,
    218
coins, and clock pendulums 220,
    221, 222
colour change, in fish 143–5
combined oral contraceptive pill,
    effects on male pill-taker 49
computers
    bubbles in nearby still mineral
        water 96–7
    and the human brain 47
    parallel/sequential processing
        38
conjunctiva 73
contact lenses 74, 134, 135
continental pole of inaccessibility
    (CPI) 153–4

contraceptive pill, effects on male
    pill-taker 48–50
'Cooee' call 210–11
corona 184
cotton
    chemically treated 82, 83, 84
    and crease retention 82
    washing before wearing 84
cremation 54–6
cricket, and Hawk-Eye 212
custard, walking across a
    swimming pool full of custard
    219
cycling, competitive, at high
    altitudes 162–4

**D**
dead bodies, disposal of
    burial 55
    consumption by vultures 55
    cremation 54–6
    donation for medical research
        55
    Promessa process 56
    Resomation 56
    in space 36–7
deafness, and loud music 69–70
death during space voyages:
    disposal of bodies 36–7
decibel (dB) scale 198, 199
dehydration
    and dark circles under eyes 31
    and downing a pint of water
        in one go 16
deindividuation 203
deodorants, men's/women's
    fragrances 30
detergents, in shampoos 90, 91
diastolic flaccidity 40
digital technology 80, 81
dinosaurs, non-jumping 106
dizziness
    and bats 150, 151
    loved by children, hated by
        adults 74–5
dogs
    ash levels in dog food 141–2

breeding for intelligence
126–7
drinking method 123
and motion sickness 193, 194
posing and threatening
lapdogs 205
temperature regulation 123–4
tolerance of cold water 122–4
dolphins
blowholes 116–18
non-drinking 107
dopamine 70
downforce 194–6
ducks, recognition of own young
112–13

E
Eagle lunar module 172
ears
ear wiggling 72–3
interpreting the acoustic
environment 84
pressure transducers 84
Earth
curvature of the 155–7
freezing after Sun
extinguished 158–61
echolocation 150, 152
elands, jumping ability 105
Electric Brae, Ayrshire 166–7
electricity, 'silent servant'
description 97, 99
elephants, jumping 104–7
emotions, affecting the heart
39–40
energy, used in cremation 54–5,
56
epichlorohydrin 83
erythrophores 143
ethanol
metabolised by the liver 52
and mood 70
as a wetting agent 10
European Southern Observatory
172
evolution 116, 131
excretion, two systems 28–9

eyes
dark circles under 31
inverted images corrected by
brain 41–3
itchy 73–4
red 73

F
faeces 28–9
false-twist textile process 91
fat
in 'female areas' 49
goose 17–19
as an insulator 38–9
and milk 137
fenugreek, burnt-sugar smell
of 20
'film formers' 91
fires
fire-lighting in cold
temperatures 176–7
ignition point 177
fish
blindfolded 143
colour change 143–5
transporting 193
flash floods 188
flatfish, colour change 143, 144
flavonoids 111
flounders, colour change 143
foods
aversion to some spices 51–2
dog foods 141–2
electrical arcing in microwave
oven 16–17
fat content 142
fatty foods of animal origin
191
grimaces in reaction to 43, 44
monodiet 3–5
'never a sweet before the
meat' principle 26–7
nutritional content 141
protein content 141–2
smells 20
football hooliganism 204
foreshortening 46

form (pressure) drag 215
formaldehyde 83
frost, not forming around trees
    182
fruit flies, tolerance to alcohol
    53
fungus, edible, nutritional value
    of 13–14

**G**
gag reflex 61, 62
galaxies 172–3
    lunar-orbiting 174
gamma-aminobutyric acid
    (GABA) 70
gas chromatography 25
geese
    adopted goslings 113
    goose fat 17–19
gelatin 218–19
Gemini space capsule 200
glasses
    inverting spectacles 42–3
    lenses falling out when
        exposed to extreme cold
        217–18
    wearing 134
global twists (gts) 93
global warming 148
glutamate receptors 70
gravimetric extraction 142
gravity, plants' gravity sensors
    121
greenhouse effect 158
grimacing, as reaction to noxious
    stimuli 43–4
ground effect 197
guar gum 219
gulls, worm baiting 114–15

**H**
hail
    duration of hailstorms 186,
        187
    layers of 186, 187–8
hair standing on end, and
    lightning clouds 180–81

Hawk-Eye automated tennis line-
    judging system 212–13
hearing loss, and noise levels
    69–70
heart, reflecting emotions 39–40
heat conduction 39
helicity 92–4
helplessness, causes diastolic
    flaccidity 40
honeysuckle, winding clockwise
    121
hops, winding clockwise 121
horses, and motion sickness 193
*How to Fossilise Your Hamster* 6
hummingbirds
    black-chinned 139
    ruby-throated 138–40
hydrothermal vents 160–61
hydroxyisohexyl 3-cyclohexene
    carboxaldehyde (Lyral) 89, 90,
    91–2
hypogeusia (altered taste) 52

**I**
ice
    and cloudiness in shaken
        martinis 7, 8
    in potato vodka martini 9
Icelandic volcanic eruptions
    (2010) 159
ichthyosaurs 116, 117–18
'impulsing escapement' 221
insects, colours of toxic insects
    109
insulation
    breakdown 98
    polar bears' feet 147
intelligence, in dogs 126–8
International Radiotelephony
    Spelling Alphabet 211
Inuit
    avoidance of scurvy 60–61
    early deaths from heart
        attacks and other cardiac
        problems 4
inverting spectacles 42–3
iodine, in cow's milk 137, 138

iridiophores 143

**J**
jelly-filled swimming pool
218–20
jumping, quadrupeds 104–7

**K**
kin discrimination 112
Kjeldahl method 141–2
knot-antiknot (k-a) pairs 93

**L**
ladybeetles 131
lake, a sloping lake 171
land reclamation 162
language
'Cooee' call 210–11
'um' and 'er' conversation
fillers 65–7
lateral geniculate nucleus 41
lather, in shampoos 90
light
diffraction of 184
refraction of 184
lightning strikes 180–81
lithium batteries 103
liver function tests 76
'lump-in-throat' oesophageal
spasms 40
lunar haloes 184
Lunar Reconnaissance Orbiter
174

**M**
magnetic fields 97–100
Magnetic Hill, Manitoba 167
magneto ignition systems 80
magnetostriction 97, 98
magnets, and monkey game 85–6
magpies, and colour coding of
milk bottle tops 109–10
Malpighian tubules 29
mammals
camouflaged 128–9
green 128–30
myopic 133–6

marine animals
lack of sweat glands 108
non-drinking 107
sources of water 107
martinis: 'shaken, not stirred' 6–9
medical misadventure, deaths
from 191
melanophores 13, 145
memory
memorising order of a pack of
cards 48
recall mechanisms 57–8
mental chronometry 208
mercury, swimming in 215–16
Mercury space capsule 200
methi *see* fenugreek
'method of loci' 48
mick (hammock) 101–2
microphones 84, 85
microsomal ethanol-oxidising
system (MEOS) 52
microwave detector 214
microwave ovens
electrical arcing in food 16–17
mobile phones in 213–14
milk
birds and colour coding of
milk bottle tops 109–10
pasteurised 137
raw 137
semi-skimmed 109
standardised for fat 137
vitamins and iodine
supplements 137–8
whole 109
yellow colour of frozen milk
14–15
minerals, in mushrooms 13
mirror-writing 72
mitochondria 139–40
mobile phones, in microwave
ovens 213–14
monodiet 3–5
mood, and alcohol 70–71
Moon
albedo of 165, 166
and rainbows 184

reflected light 165
  telescopic evidence of moon
    landings 172–4
morning glory plants, winding
  counter-clockwise 121
mosquitoes, transmitting
  *Plasmodium* 149
motion sickness 150, 151
  in animals 193–4
motor racing, weather forecasting
  175
motorcycles, interfering with
  digital TV reception 80, 81
muktuk 60–61
mushrooms, nutritional value of
  13–14
music written in 4/4 time,
  apparent preference for 32–4
mustard, hotness of 11, 12
myopia 133–6
myrosinase, and hotness of
  mustard 12

**N**
names
  choosing dog names 211
  remembering 57, 58
natural selection 113, 117, 123,
  126
Netherlands, total area of water
  in 161–2
nickel-cadmium batteries 102
nickel-metal hydride (NiMH)
  batteries 102, 103
nipples, male 67–8
noise levels
  Formula 1 cars 197–9
  and hearing loss 69–70
  in the workplace 69
  *see also* sound
nose picking 34–6
noxious stimuli, reactions to
  43–4
nutrients in diet 3–5

**O**
occipital (visual) cortex 41

oestrogens, in combined oral
  contraceptive pill 49
oils, in shampoos 91
online behaviour 203–4
optic nerve 41
optical illusion, rolling uphill
  166–71
orange squash, downing a pint in
  one go 15–16
otolith organs 150
oxygen
  in a biosphere 145, 146
  at high altitudes 162, 163, 164
  in peppers 24, 25–6

**P**
pack of cards, memorising order 48
Pakicetus 117
panting 123
papillae, under a polar bear's
  paw 147–8
passerines 139
pencil line length 94–5
pendulums 220–23
penile erection 68
peppers, gas composition 24–6
photosynthesis
  in a biosphere 146, 159
  in peppers 25–6
plants
  C4 145, 146
  gravity sensors 121
  winding clockwise (CW) 121
  winding counter-clockwise
    (CCW) 121
*Plasmodium* 149
plastics, coefficient of expansion
  217, 218
polar bears, insulation 147–8
Polaris missile 153–4
polders 162
possum, green ringtail 129–30
presbyopia 136
preservatives, in shampoos 91
progestogen-only oral
  contraceptive pill, effects on
  male pill-taker 49

Promessa 56
proprioceptors 75

R
racehorses, jumping ability 105
radiation, and snow melting
    faster under a tree 182
rain, running to reduce exposure
    to rain 185–6
rainbow, U-shaped 183–4
Raleigh criterion 173
ramps, energy used in climbing
    202
rats, plague-carrying 148, 149
reaction times, measuring 208–9
recall mechanisms 57–8
rechargeable batteries 102–3
recognition reaction time
    experiments 209
reduced cues theory 203
reflection 183
Resomation 56
respiration, rate in peppers 25
retina
    and colour change in fish
        143, 144
    converts image into neural
        information 41, 42
rhinotillexomania 35
rhythmic choice 34–6
riboflavin, colouring frozen milk
    yellow 14–15
rock music 69
rolling uphill 166–71
ruby-throated hummingbird
    138–40
running, competitive, at high
    altitudes 162–4

S
scooters, interfering with digital
    TV reception 80
scurvy 4, 60
seawater, density of 22
shampoo ingredients 89–92
sinigrin, and hotness of mustard
    12

sirenians 116
sleep
    not falling out of bed 100–101
    sleep in a mick 101–2
sleepiness, and dark circles under
    eyes 31
slo (streptolysin O) gene 53
sloth, three-toed 128, 129
smells
    and birds recognising each
        other 112
    dank 87–8
    food 20
    shampoos 90, 91–2
snow
    duration of snowstorms 186–7
    layers of 186, 187
    melting faster under a tree
        182
snowflakes 186
sodium diethylene-triamine
    pentamethylene phosphonate
    89
Solanaceae 110
solar haloes 184
solute concentration, and
    floating/sinking citrus slices
    23–4
sotolon, powerful smell of 20
sound
    background noise 84–5
    and birds' recognition of own
        young 112–13
    refraction 178, 179
    speed of 178
    traffic noise 178–9
    see also noise levels
sound waves 178, 179
space voyages, disposal of body
    after death during 36–7
spacecraft, orientation on
    returning to Earth 199–201
spectacles see glasses
spices, aversion to some spices
    51–2
spirit levels 171
spleen, of dogs 124

Spook Hill, Lake Wales, Florida
169
stair-climbing, energy used in 202
starting guns 208, 209
Stefan-Boltzmann law 158
sticky tape, stress concentration
78–9
stomach aches, due to anxiety 40
stress concentration 78–9
sugar, 'never a sweet before the
meat' principle 26–7
Sun
freezing of the Earth after its
demise 158–61
nuclear fusion 160
and rainbows 183–4
surface tension, and buoyancy of
appletini garnish 10, 11
swimming
in guar gum 219
in liquids other than water
215–16
pressure-sensitive wall pads
205–7
sword-swallowing 61–4

T
Tambora volcano eruption (1815)
158
taste
altered 42
aversion to some spices 51–2
Teflon resins 83
teleost group 143
telephone cords, knots in 92, 93–4
telescopes
Gran Telescopio Canarias
172, 173
Hubble Space Telescope 174
resolving power 172, 173
telescopic evidence of moon
landings 172–4
television, motorbikes and
scooters interfere with
reception 80–81
temperature
air 125

comfort level of bath vs. room
temperature 64
dogs' temperature regulation
123–4
overnight cooling 159
tennis, Hawk-Eye automated
line-judging system 212–13
testosterone, and the
contraceptive pill 49, 50
*Theileria* parasite 149
thiamine, yeasty smell of 20
thickeners, in shampoos 91
thinking about actions, and
automatic activity 37–8
time signatures 32–4
tinnitus 69
tiredness
and blood vessels in eyes 73–4
and dark circles under eyes 31
tomatoes, sticky fluid on leaves
110–12
towels
cut-pile side 83
drying qualities 82–3
looped terry side 83
transformers
hum of 97–9
silent 99–100
transsexual people, and brain
volume 49
truffles, nutritional value of 13
tube worms 160
tubulin 121
typing, automatic 37

U
'um' and 'er' conversation fillers
65–7
Umbelliferae family 51
urine 28, 29

V
vestibular system 75
Vietnam War 190–91
vines, and winding direction 121
virtual worlds 204
viscous drag 215

vision
    binocular 72
    close objects apparently
        moving faster than those
        further away 44–7
    inverted images corrected by
        brain 41–3
    perfect (emmetropia) 133
    slowness of 151–2
vitamins 4, 5
    in cow's milk 137–8
    and a monodiet 4, 5
    in mushrooms 13, 14
    vitamin C in Inuit diet 60–61
vodka martinis, 'shaken, not
    stirred' 6–9
vomiting, and motion sickness
    193, 194
voodoo curse 40

W
water
    bubbles in still mineral water
        96–7
    cats' intolerance of 122–3
    density of carbonated/non-
        carbonated water 21–2
    dogs' tolerance of cold water
        122–4
    downing a pint in one go
        15, 16
    flowing uphill 170–71
    marine animals' intake 107,
        108
    and smelly items 87–8
    total area of water in
        Netherlands 161–2

waterfowl, inability to keep track
    of own young 113
wave-making drag 215–16
weaponry, fatalities 189, 190
weapons systems, and
    the continental point of
    inaccessibility 153–4
weather radar 175–6
whales
    baleen 117
    blowholes 116–18
    blue 117
    humpback 117
    non-drinking 107
    sperm 117
wind shear phenomenon 178
wind speed, and traffic noise
    178–9
wisteria, winding counter-
    clockwise 121
wolves, tolerance of cold water
    122
worm baiting 114–15

X
xanthophores 143

Y
Yersinia (plague microbe) 149
yodelling 211

Z
zinc deficiency, and altered taste
    42